{ MOMENTS OF TRUTH
IN GENETIC MEDICINE }

{ MOMENTS OF TRUTH
IN GENETIC MEDICINE }

Susan Lindee

THE JOHNS HOPKINS UNIVERSITY PRESS
Baltimore

© 2005 The Johns Hopkins University Press
All rights reserved. Published 2005
Printed in the United States of America on acid-free paper
2 4 6 8 9 7 5 3 1

The Johns Hopkins University Press
2715 North Charles Street
Baltimore, Maryland 21218-4363
www.press.jhu.edu

Library of Congress Cataloging-in-Publication Data

Lindee, M. Susan.
Moments of truth in genetic medicine / Susan Lindee.
p. cm.
Includes bibliographical references and index.
ISBN 0-8018-8175-7 (hardcover : alk. paper)
1. Medical genetics—United States—History. I. Title.
RB155.L555 2005
616'.042—dc22
2004028267

A catalog record for this book is available from the British Library.

To Grant and Travis

CONTENTS

Acknowledgments / *ix*

CHAPTER 1
INTRODUCTION / 1

CHAPTER 2
BABIES' BLOOD
Phenylketonuria and the Rise of Public Health Genetics / 28

CHAPTER 3
PROVENANCE AND THE PEDIGREE
Victor McKusick's Field Work with the Pennsylvania Amish / 58

CHAPTER 4
SQUASHED SPIDERS
Standardizing the Human Chromosomes and
Other Unruly Things / 90

CHAPTER 5
TWO PEAS IN A POD
Twin Science and the Rise of Human Behavior Genetics / 120

CHAPTER 6
JEWISH GENES
History, Emotion, and Familial Dysautonomia / 156

CHAPTER 7
CONCLUSIONS / 188

Notes / *211*
Essay on Sources / *231*
Bibliography / *237*
Index / *261*

CONTENTS

[ACKNOWLEDGMENTS]

My first and most profound debt is to Dorothy Nelkin. Dot was my mentor, coauthor, and friend. She read and commented on most of what I have written, not only for this project but for almost everything I have worked on over the past two decades. She was involved with this study from its earliest stages to the near-final manuscript, reading many chapters and providing detailed suggestions and comments. Her clear-eyed assessments were invaluable, her judgments impeccable. Her courage throughout her life was a gift to all who knew her. For me, many times, she held up the sky, and I will miss her always.

My colleagues in the Department of the History and Sociology of Science at the University of Pennsylvania have been supportive, tolerant, skeptical, and engaged. Conversations, observations, criticisms, and feedback from Robert Aronowitz, Ruth Cowan, Steve Feierman, Riki Kuklick, and Janet Tighe have been particularly helpful.

Charles Rosenberg, back in the old days when he was still at Penn, read several versions of some chapters and suggested many improvements. Then, after he had departed for Harvard, he carefully read the entire manuscript in near-final form and proposed a shift of perspective that greatly improved the final version. He has encouraged my work on this project throughout its long gestation, and my debts to Charles over the years are profound.

Graduate students (past and present) at Penn, including Josh Berson, Paul Burnett, Eve Buckley, Betsy Hanson, Andi Johnson, Joanna Kempner, Susan Miller, Chloe Silverman, Jeff Tang, Dominique Tobbell, Roger Turner, and Audra Wolfe, provided insights, support, teaching assistance, and encouragement on many levels. Chloe Silverman's work on autism was a constant source of inspiration, helping me think about patients and biomedicine in productive ways. Jon Merz, at the Penn Center for Bioethics, kept me aware of alternative ways of interrogating contemporary genomics, and through the generosity of Art Caplan and the Penn Center for Bioethics I had a quiet place to work during a sabbatical leave.

Other colleagues politely listened to me complain and fret, including Mark Adams, David Barnes, Robert Kohler, Nathan Ensmenger, and Nathan Sivin. Patricia Johnson provided administrative and social support.

At Johns Hopkins, Barton Childs corrected and inspired me and was patient with my questions. I suspect that Barton is a few decades (centuries?) ahead of his time, and it was a great privilege to be able to draw on his insights. Scott Gilbert, at Swarthmore College, read chapters, provided feedback, and laughed at the jokes. He and Barbara Kimmelman, at Philadelphia University, also met me for bagels and conversation once a month during a particularly confused period early in the project.

I benefited from ongoing feedback from Rayna Rapp, whose commitment to justice I greatly admire, and from Vassiliki Betty Smocovitis, who kept this project on track at several points. Faye Ginsburg was an informant, commentator, and critical supporter, sending website addresses, e-mail threads, amazing quotes, and sometimes T-shirts. Karen-Sue Taussig provided commentary, criticism, encouragement, and occasional dinners out.

I am also deeply indebted to Diane Paul and Jonathan Marks, both of whom exemplify critically engaged scholarship.

Some people who played a role in the events I explore read and commented on chapters or spoke with me about their research and their fields. Barton Childs helped me understand the rise of PKU testing. John Hamerton's comments on the cytogenetics chapter were perceptive and extremely helpful. Clarke Fraser generously read and commented on every chapter, caught many infelicities, and encouraged and praised what was worthwhile. I am deeply indebted to him; he was an excellent critic in every way. I also drew on the insights of Felicia Axelrod, Joseph Dancis, Irving Gottesman,

Gladys Gropper, Alfred Knudson, Barbara Migeon, William Pollin, Artemis Simopoulis, and Conrad Riley.

Thanks are due to Joyce and Richard Rosen, who happen to live across the street from the house where I grew up in Houston. Joyce and Richard, who have an affected son, told me about familial dysautonomia, a genetic disease I had never heard of, and this disorder subsequently became the focus of an important chapter in this book.

Archivists are always a crucial resource, and I am grateful to Daniel Barbiero and Janice Goldblum at the National Academy of Sciences; Marjorie Ciarlante and Tab Lewis at the National Archives and Records Administration at College Park; and Matt Fulghum at the National Archives and Records Administration in Washington, D.C. Special thanks are due to archivists Nancy McCall, for her guidance and suggestions, and Gerard Schorb, for his patient and efficient assistance in my efforts to negotiate the papers of Victor A. McKusick at the Alan Mason Chesney Medical Archives at Johns Hopkins University. Thanks are due to McKusick, for his permission to use this remarkable, rich collection.

Marion Robertson, at the Maryland Department of Health, is neither an archivist nor a historian, but she saved old records in an unused cabinet in Baltimore and I am grateful for her foresight.

The Burroughs Wellcome Fund Fortieth Anniversary Award supported this project, providing funds for crucial writing and research time. The award permitted me to begin serious research on a project that had been quietly germinating for some time. I only regret that its completion has take me so long, despite the fund's support and that of the University of Pennsylvania, which provided me with sabbaticals over the years.

Jackie Wehmueller, at the Johns Hopkins University Press, was a sympathetic and enthusiastic editor. Linda Strange copyedited a final manuscript that desperately needed her skills. Paul Burnett patiently tracked down permissions for the illustrations.

Permissions to reprint earlier published works were graciously given by the Chicago Kent Law Review, for a version of chapter 2 published as "Babies' Blood: Phenylketonuria and the Rise of Neonatal Testing, 1955–1965" (*Chicago Kent Law Review* 75, no. 1 [1999]: 113–133); and also by the University of California Press, for a version of chapter 3 published as "Provenance and the Pedigree: Victor McKusick's Fieldwork with Ellis–van Creveld Syndrome in the Pennsylvania Amish" (in *Genetic Nature/Culture: Anthropology and Science beyond the Two Cultures Divide*, ed. Alan Goodman, Deborah Heath, and Susan Lindee [Berkeley and Los Angeles: University of California Press, 2003]).

On a personal level, thanks are due to Dot's husband, Mark Nelkin, and daughter, Lisa Nelkin. I have also relied heavily on the generous support of my sisters, Marguerite Lindee, of Houston, and Lauren Lindee, of Tregarth, Wales; my brothers, Michael, Herbert, and Charles Lindee, all of Houston; and my friends Annie Dannenberg of Arcata, California; Lee Dante of Philadelphia; Barbara Kimmelman of Philadelphia; Sydney Rubin of Austin, Texas; and Betty Smocovitis of Gainesville, Florida. In this group, only Betty and Lauren read chapters, but all helped me produce this book nonetheless. The caring labor they performed may not count as standard scholarly assistance but it mattered a lot to me.

Finally, I wish to remember with gratitude my husband, Brett Skakun, whose struggle with cancer ended in the fall of 2003. Brett supported my work from its earliest stages. This book is dedicated in his memory to our sons, Grant and Travis Skakun, with love.

CHAPTER I

INTRODUCTION

Human genetics was transformed from a medical backwater to an appealing medical research frontier between 1955 and 1975. This periodization was not obvious to me when I began my study, but over and over again I found these twenty years emerging as a turning point in the fortunes of genetic disease. This book, therefore, explores the institutions, disciplines, practices, and ideas that began to reconfigure human disease in genetic terms during what might be called the long 1960s.

For many critical intellectual and institutional innovations, the crucial period was even shorter. There were remarkable changes in many fields relevant to genetic disease during the four or five years after 1959. For public health genetics, behavior genetics, cancer genetics, and biochemical genetics, the early 1960s were transformative. Human cytogenetics was a sleepy subspecialty of no interest to physicians in the late 1950s. By 1964 it was glamorous enough that practicing clinicians wanted to learn to work with chromosomes. Public health genetics did not exist, at least in the sense of legislative or political support, in 1960. But by 1966 most U.S. states had elaborate neonatal testing programs created by legislators intrigued by phenylketonuria testing and clamoring for more tests. In the early 1960s, experts interested in genetic disease (and they were not all geneticists by any means) seem to have hit something like criticality—in physics, the point at which there is a sufficient quantity of fissionable material to sustain a chain reaction, and in culture, perhaps, the point at which interactive effects produce rapid institutional and social change. There were enough people and institutions with a stake in genetic disease to support a series of related, and sometimes independent, events that have had profound consequences for the development of biomedicine. And so this book is a study of a period of transformation in one of the most high-profile biomedical fields of the late twentieth century.

At another level, it is a study of the realization of an idea. The idea is that all human disease is a genetic phenomenon subject to technological control. The realization of this idea is its tangible manifestations, which are threaded through the practices and policies of institutions and through the most technical and most intimate interactions—between parent and child, physician and patient, or scientist and research subject. The idea that all disease is genetic disease is not an abstraction. It is a social experience manifest in language, technology, emotion, and policy. Increasingly, it plays a role in the legal system, in the practices of hospitals, in the research priorities of the armed forces, in the treatment of people with mental retardation, and in medical education. It has become an idea with social force. The study of the realization of an idea is the study of the imbrication or embeddedness of that idea and the study of how and why it works and what it accomplishes. It is also the study of the many diverse workers involved in building the network of policies and practices that hold the idea in place and mediate its consequences. When I look at the events that form the core of my story, then, I am looking for the idea that all disease is genetic disease as it solidified across so many social and epistemological fields and as it became one of the most important ways of understanding the frailties of the human body.

Finally, at a more theoretical level, this book is a study of the patchwork qualities of knowing. The "moment of truth" of my title is the moment of recognition or understanding, the moment when a given phenomenon is classified or categorized or placed in a narrative that explains it. It can be a moment experienced by anyone—the scientist, the research subject, the families of the research subject, or even people in general. In my construction, it becomes a moment of truth by virtue of having been incorporated into a formal scientific text or into scientific and medical practice. At that point, it has been taken up as a factual detail, a scrap of reality, perhaps with some of the edges trimmed so that it is in effect camouflaged, presented seamlessly along with other moments of truth experienced by other actors.

This is a counterintuitive way of seeing scientific texts, which are usually understood to be products of the knowledge of the author or authors or perhaps of the scientific community that polices publication. In some ways, scientific papers are precisely that, produced by a single person or laboratory or research group and expressing the negotiated knowledge of that person, laboratory, or group, as filtered through the broader scientific community. But in other ways they are sometimes amalgamations of knowledge acquired by many different people at many different times. What this perspective contributes to our understanding of biomedicine is one of my central problematics, and the problem is by no means resolved here. At the very least, it suggests that the process of producing technical knowledge is even more complicated than it seems. Scientific knowledge is a community project that can be (has been?) shaped quite directly by people in many different social and professional locations. Although professional authority looms large in many accounts of technical knowledge, the power to experience and report truth may be much more dispersed than it sometimes appears.

A moment's reflection suggests how active human beings are as subjects in an experimental system. Those who work with mice, flies, or other nonhuman organisms do encounter actions that might be identified as resistance in their subjects. Flies die or fail to breed; mice behave in unexpected ways. Organisms have biological agency that is the direct source of their value to science, and when they resist, when they perform in unexpected ways, their resistance is informative and revealing. Their agency is what makes them potential resources for the elucidation of natural truth. But human beings have a form of agency that is both more powerful and more troubling. They have what might be called contaminated agency. Like scientists, human subjects are sentient and capable of ignorance, emotionality, superstition, and malice. A mouse cannot be unreliable in this way, but human beings are commonly understood to be unreliable in this way at all times. Human subjects are capable of resistance, hostility, illogical thinking, and dishonesty. At the same time, they can also be extremely informative and insightful, keen observers of their own world and their own bodies, able to illuminate technical problems by virtue of their historical and social experiences.

About two years into this project, I began to notice that conclusions about heredity and disease reached by people who were not scientists and not technically trained were sometimes incorporated whole into scientific texts.

I noticed, too, that distinctions between different forms of knowledge were often flattened out or elided or disappeared in such texts. I became interested in the presence of folk knowledge, emotional knowledge, craft knowledge, or social knowledge in these kinds of technical accounts, and I paid attention to whose experiences and ideas were inside the scientific texts or guiding the scientific projects, shaping the technical questions and expected solutions. For many of my actors, including physicians, geneticists, and other scientists, knowledge generated by people who were not technically trained was not interpreted as corrupted or polluted knowledge. It was instead a familiar, usable, productive type of information, acquired first-hand by persons who had plausible and perhaps trustworthy reasons for their conclusions.

In some cases, folk or social knowledge seemed to be an unproblematic part of the intellectual architecture of the discipline or subfield. Behavior geneticists, for example, attributed certain beliefs to the common man and then suggested that such widely held beliefs validated their own scientific results: everyone already knows personality is inherited, they proposed, so their technical conclusions must be right. In other cases, social experience provided informants with knowledge so trustworthy it could be used as a surrogate for established methods of technical analysis. Since the 1960s, twins themselves have been the primary source of information about their own zygosity. Twins tell geneticists whether they are identical or fraternal. Based on social experience, they know whether they came from one egg or two eggs. They can condense a lifetime of reactions to them by friends and family members into a yes or no answer to the question, "Were you and your twin as alike as two peas in a pod?" And their condensation of this social experience has consistently proven reliable when compared with blood tests and phenotypic analysis. The "questionnaire method of determining zygosity" became and remains the primary method of zygosity diagnosis in twins research.[1] The moment of truth, for this diagnosis, is experienced by the twins. It is theirs. They are the reliable knowers.

In another example, the Pennsylvania Old Order Amish have been active collaborators in some of the genetic research conducted in their community. When the Johns Hopkins University medical geneticist Victor McKusick began working with the Amish in the 1960s, he found a deep local narrative of disease and heredity that permitted him to construct lengthy family pedigrees. These pedigrees depended on notes in Bibles, reports from the local undertaker, memories of midwives and mothers of infants long dead, testimony from fathers and uncles, state birth and death records, and other sources equally varied. The moments of truth—of recognition, interpretation, understanding—in each pedigree were experienced by many different people, in different times and contexts, and were recorded or remembered in many different ways. McKusick's published papers mention, in appendices or notes, conversations in the local market or chance encounters in a hospital corridor, where people in a position to know told him things about disease, babies who had died, or cousins who had extra fingers.

Similarly, physicians baffled by a rare genetic disease, familial dysautonomia, drew on parental knowledge and experience to understand it, partly because the disease itself made it more difficult for them to see what was happening. Physicians commonly saw cases of familial dysautonomia only when the child was in autonomic crisis. Just the stress of coming to the hospital could bring on intense vomiting. But parents had daily experience with such children, and in the years shortly after the disease was first described, physicians asked parents to meet with them every month and to describe, in these meetings, what their children did when they were calm. Mothers and fathers knew things about these complicated, puzzling young bodies with which they interacted so closely. Parents saw things about genetic disease. They had knowledge. Perhaps they even had the truth.

In the published scientific papers I read for this study, in genetics journals of many kinds, sometimes the presence of these varied voices and perspectives was explicit. And sometimes the presence of these voices and perspectives was not visible within the text but could be tracked by looking at the records of field research or private correspondence. The knowledge of research subjects—the things they knew about themselves and their world—could sometimes be seamlessly taken up and blended in, its personal origins elided. The patchwork thus produced was a scientific paper that depended on the experiences and interpretations of many people, without necessarily seeming to do so.

I find the textile analogy of the patchwork quilt apt in several ways. Quilts are folk art and products of common knowledge. They are low art, not high, just as social experience or emotional knowledge is low knowledge, not high. The skill of quilting is, furthermore, the special province of a social group traditionally excluded from science. Women's bodies, for much of the history of Western science, made them unreliable knowers and untrustworthy witnesses to nature's ways. So, too, the social structure of science has, since the professionalization of science in the nineteenth century, carefully excluded nonscientists from the ranks of reliable knowers. The proper scientist, with a Ph.D., a history of publications, and an academic or industrial post, has the socially agreed-upon expertise and authority to explain nature. These explanations might at times mirror folk belief, but the status of the person providing them makes them scientific.

Those who are outside this authoritative structure, however, can and do come to know important things about natural phenomena—by experience, observation, shrewd logic, or shared culture. I sometimes ask my students to choose who they would consult if they needed to grow a field of beets. Their first option is a scientist at the local U.S. Department of Agriculture experiment station, who has deep technical knowledge of beet genetics, beet biochemistry, and the evolution of beets. Their second option is a semiretired beet farmer, who has been growing beets for forty years. They pick the beet farmer. Even students fully socialized in the values of a technocratic society think the beet farmer knows important things. In my own construction, the imagined quilters referenced earlier are symbolic of those who do not have the technical authority to speak for nature but are nonetheless understood to know important things—farmers, but also patients, parents, research subjects, the public.

Finally, the patchwork quilt is a coherent whole derived from many sources, a pattern incorporating bits of someone's fraying shirt and someone else's outgrown skirt. The origins of some quilts might be meaningful—for example, one made for a particular person from all her baby clothes. The origins of other quilts might be irrelevant, subservient to the desired pattern. Like the scientific paper, then, the quilt can highlight or downplay origins. The fabric can come from many sources. But those sources might or might not be important enough to give meaning to either text or quilt.

I suggest here that knowledge of many different kinds, of different provenances, was incorporated into the formal scientific record of genetic disease. And by implication, I suggest that biomedical knowledge more generally expresses community knowledge in ways we have barely considered in our theorizing in science studies.

In an earlier study, with Dorothy Nelkin, I looked at images of DNA in popular culture. We explored the different ways in which genes appeared in advertising, soap operas, comic books, and books of popular advice, noting how DNA moved across cultural and political fields, so that cars and magazines could have genes and judges could invoke heredity to explain legal decisions. We were interested in the free fall of scientific ideas, concepts, and explanations "down" into popular culture (Nelkin and Lindee 1995). But we were also implicitly exploring the traffic in the other direction: how did popular notions work their way into technical conclusions and research programs? We were suggesting that nonobvious values and expectations crucial to science as a technical enterprise were expressed overtly in popular texts. We were also delineating a rough feedback loop, in which technical experts drew on common knowledge and popular expectations as they fashioned their studies (Marks 2002).

This book follows from some of those earlier questions, using different sources and framing. Methodologically, I started with the technical knowledge and then tried to track it, to see who contributed what, and to notice who personally experienced the moment of truth that appeared in the published scientific text, shaped public policy, or produced clinical practice. Genetic disease is a particularly rich focus for such a study because folk or practical knowledge of heredity long predates scientific knowledge.

The simple observation that like begets like was presumably preliterate.[2] By the mid-nineteenth century, when Charles Darwin developed a theoretical frame to explain biological change through time, a vast, sophisticated corpus of knowledge based on domestic breeding of useful animals was at his disposal.[3] The controlled reproduction of corn, chickens, sheep, or cattle required skills that were culturally inherited, learned by each new generation from its elders. These practices and strategies in turn reflected the biological qualities of the organisms being bred. Pig breeders knew pig behavior, reproduction, temperament, and growth because these qualities were important to successful breeding, and this knowledge shaped practices of management. Breeding was and is complicated biological work, requiring deep knowledge of the organism itself.[4] In some cases, when the value of the organism depends on lineage, as with racehorses and purebred dogs, domestic breeding involves elaborate pedigrees, which were originally developed as systems for attesting to the quality and authenticity of the animal (Ritvo 1987). Darwin and later scientists interested in heredity and evolution were thus building on practical breeding knowledge and techniques of long human experience.

Similarly, the existence of diseases that were "in the blood" was recognized long before anything resembling scientific human genetics came into being. The notion of constitutional illness has a long history in medical thinking and practice. Nineteenth-century physicians and patients understood illness to reflect the body's inborn traits, adopting a broad notion of what could be considered inborn. Parental moral failings, for example, could be manifest in the constitutional weaknesses of offspring. As Charles Rosenberg noted in a widely cited essay on constitutional disease, "The sins of the fathers could thus act literally as a plague debilitating and stigmatizing their children" (Rosenberg 1997, 33).

A few specific hereditary diseases, such as hemophilia, were discussed in medical texts in the medieval period, and gross malformations of babies born with mermaid-like legs or single eyes are depicted even earlier: the malformed body is the most obvious kind of congenital, not necessarily genetic, disease (Daston and Park 1998). But in the nineteenth century, many more conditions that eventually were understood as genetic were described and named. Down syndrome was identified as mongolism in 1866 by the British physician John Langdon Down, who thought he was seeing throwbacks to the Mongol race; Huntington disease was described as a familial syndrome in 1872 by the American physician George Huntington.[5] Tourette syndrome was described in 1885 by the French neurologist Georges Gilles de la Tourette, though it was not understood as genetic; osteogenesis was described in several stages from 1835 to 1918; and xeroderma pigmentosum in 1874. Some of this diagnostic fervor was a product of the changing status of physicians and their increasing professionalization. Many new diseases were named and described in this period. And the status of some of these diseases as genetic was contested or unclear. But by 1900, genetic disease was not wholly novel. It was an established biosocial reality. Physicians could recognize some genetic diseases; farmers and breeders knew that genetic diseases appeared sometimes in their stock; and more generally, the commonplace observation of familial resemblance was the subject of folk interpretations of behaviors and differences.

In the American eugenics movement, such folk interpretations appeared in the attribution of complex mental, social, and political problems to heredity. Eugenics promised a better future that would depend on better germ plasm. Guaranteeing these improvements in the germ plasm would generally, at least under American law, require involuntary sterilization of those who were unfit. This was the basis of the involuntary sterilization laws passed in so many states in the 1910s and 1920s.

MOMENTS OF TRUTH IN GENETIC MEDICINE

Some advocates of eugenics were virulent racists; others were ministers, popular science writers, feminists, politicians, or health care advocates; still others were trained in zoology, medicine, or natural history. Practicing geneticists around the world were also interested in eugenics and its potential to improve the human condition and to alleviate human suffering. The popularity of eugenics in Britain, the United States, Canada, Brazil, Japan, France, Germany, and other nations suggests its broad plausibility and appeal. Its agenda of improving the human germ plasm through political restrictions and propaganda justified sustained research on human heredity. Some of that research, grounded in the hope of improving the human species just as horses or cattle had been improved, led to novel technical insights. Eugenics provided a deep structure or a foundational idea for serious technical work. It also provided a foundational idea for Nazi racial hygiene, of course, and for genocide. Primarily because of the Holocaust, eugenics was and remains a complicated problem for practitioners of human genetics and, now, genomic medicine.

One important way to manage the eugenics legacy has been to invoke a different founding narrative that places genomics not in the line descending from eugenics but in that beginning with Archibald Garrod's 1909 book, *Inborn Errors of Metabolism*. As Barton Childs pointed out in a 1970 paper in the *New England Journal of Medicine*, Garrod's citation record improved dramatically in the 1960s. From July 1964 to July 1968, he was cited 137 times by 88 different authors publishing papers on genetic disease, though admittedly most of these references "seem perfunctory . . . and are so stereotyped as to suggest that [Garrod's] papers are more often quoted than read" (Childs 1970, 72). Perhaps Garrod's perceptive insights into biochemical individuality were not the primary justification for citing him.

Yet Garrod was genuinely in tune with the intellectual strands that came together in the 1960s. His ideas did not resonate with his contemporaries in the early twentieth century, but they seemed deeply meaningful to a later generation that sought to understand heredity in biochemical terms. A prominent and well-respected British physician, Garrod began to write about chemical individuality in 1899 and made the subject a focus of his research for the rest of his life. He explored the subject in twenty-five or thirty papers, two books, and a textbook of pediatrics (Childs 1970, 72). He understood that these inborn errors were hereditary, and he believed that there might be many more than had yet been described, the human body engaging in so many complex processes that there was "room for an almost countless variety of such sports" (Garrod, quoted in Childs 1970, 72). Every individual might have a unique chemistry, he proposed, and many of these variations might be clinically invisible because they had no real impact on health. Garrod furthermore equated constitution or diathesis with these chemical differences. He localized a longstanding, extremely vague medical notion (constitution) in the body's metabolic processes and even attributed variations in metabolism to chromosomal difference. Garrod also proposed that mutations could be both good and bad at the same time, an idea, as Childs points out, that was "surprisingly modern" and was fully validated in the 1950s with the elucidation of the complex relationship between malarial environments, sickle cell trait, and sickle cell disease.

When first published, however, Garrod's ideas were not recognized as revelations by either geneticists or physicians. Reviews of Garrod's book mentioned the details of the diseases he discussed and ignored the ideas that he was using these diseases to explore. "That his name would be known to every medical student 30 years later, none of his contemporaries would have predicted" (Childs 1970, 74). He came into his audience posthumously with the rise of biochemical genetics in the 1950s. He also then became a founding father for human biochemical genetics.

Human genetics as a scientific and medical field crystallized around 1959. In that year, Harry Harris published his textbook *Human Biochemical Genetics;* French pediatrician Jerome Lejeune found an extra chromosome in children with Down syndrome; McKusick began planning the first Bar Harbor short course in medical genetics for health care professionals; and federal maternal and child health funds were first allocated to testing and dietary therapy for phenylketonuria. Garrod, the clinician with an interest in biochemistry, was the proper founding figure for this technically driven enterprise, with its amalgamation of public health, sophisticated laboratory work, and medical education.

Convincing physicians that they should care about genetic disease was one of the top priorities for a generation of tenuously professionalized human geneticists after 1945. As the fly geneticist, eugenicist, and Nobel Laureate H. J. Muller put it in 1949, the group "hoped to be able to avoid that dilettantism which has in the past characterized so many attempts to study human heredity" (Muller 1949, 1). The 150 founding members of the American Society of Human Genetics in 1948 included *Drosophila* geneticists, physicians, educators, psychiatrists, anthropologists, sociologists, and zoologists. The only properly pedigreed founding father was probably the University of Michigan geneticist James V. Neel, then just beginning his career with both an M.D. and a Ph.D. in fly genetics. Unlike many in this group, Neel was both credentialed and self-identified as a student of human genetics. He published a landmark paper on sickle cell anemia in 1949. Muller was not an M.D., and his research focused on flies and radiation. Charles Cotterman, first editor of the *American Journal of Human Genetics*, established in 1949, was trained in zoology and entomology (Cattell 1955).

Cotterman's first editorial board was similarly eclectic. Madge Macklin was a physician at Ohio State University who had done some genetic counseling (Cattell 1955). Bronson Price was a government statistician with a Ph.D. in psychology and biometrics. He had worked in the USSR in the 1930s, and he had published research on twins (Cattell 1972). Alexander Weiner was a Brooklyn hematologist.[6] C. Nash Herndon was a physician at Bowman Gray School of Medicine at Wake Forest University and, in 1952–55, president of the American Eugenics Society. When Herndon gave up that presidency in 1955, he immediately became president of the American Society of Human Genetics, thereby constituting one of the many links between eugenics and postwar human genetics (Cattell 1955; Paul 1998). Norma Ford Walker, working with genetic disease at the Hospital for Sick Children in Toronto, had a Ph.D. in biology. Walker was interested in multiple births, taste reactions, dermatoglyphics, mental defects, and genetic susceptibility to polio (Miller 1999; Cattell 1955). Horace Norton had a Ph.D. in mathematics and was a professor of statistics at the University of Illinois at Urbana. He taught eugenics in the 1930s, and later he worked as a statistician for the U.S. Atomic Energy Commission (Cattell 1972). The range of expertise and interests suggests that human genetics was multidisciplinary from the beginning, with its embrace of both high reductionism and complex clinical care. Human genetics as a nascent discipline liberally embraced technological optimism, for otherwise its difficulties seemed insurmountable.

This group of aspiring and sometimes actual human geneticists faced a population of clinicians who generally saw genetic disease as limited to rare, often bizarre conditions they would be unlikely to encounter in clinical care. Constitution, which included but was not limited to heredity, was still expected to play a role in all disease states, but genetic disease was, say, hemophilia, not high blood pressure or a compromised immune system. In Edward Yoxen's evocative phrase, genetic disease was a "medical ghetto" (Yoxen 1982). Medical schools did not offer full courses on genetics, though, after 1950, many textbooks mentioned a few well-known genetic diseases. The idea that heredity might explain virtually all diseases was not a part of medical culture. The most arresting model for disease was still infectious disease, as it had been since the 1880s and the rise of the germ theory.

This began to change in the early 1960s. Health care professionals were attracted particularly to cytogenetics, in the wake of the chromosomal explanation of Down syndrome. Reports about other chromosomal diseases rapidly followed Lejeune, Gautier, and Turpin's 1959 paper (1959a), linking abnormal chromosomes to other forms of mental retardation, congenital anomalies, and sex anomalies. These were practical discoveries. They connected easy-to-understand cellular differences to complicated bodily states. Physicians, if medical publications are any measure, were fascinated, and cytogenetics enjoyed a decade of phenomenal visibility.

Meanwhile, David Hungerford and Peter Nowell's study of chromosomes in leukocytes from the blood of children with leukemia suggested a connection between chromosomes, genes, and cancer that galvanized several scientific communities. Nowell was a former Navy pathologist interested in leukemia, Hungerford was a graduate student skilled at cytogenetic preparation, and their collaboration at the University of Pennsylvania led to the demonstration of breakage in chromosome 22 in leukemic cells. This was the first observation of a cancer-specific somatic genetic change. Admiring colleagues dubbed the shortened 22 the "Philadelphia chromosome," a nickname that persists into the present (Hungerford 1961; Freireich and Kantarjian 1993).

On another front, the discovery of an effective dietary intervention for the rare biochemical disorder phenylketonuria made genetic disease a public health problem in a new, pressing way. Children born with phenylketonuria, or PKU, developed pale white hair and severe mental retardation by about eighteen months of age. In the late 1950s physicians in Britain and the United States began exploring the possibility of dietary restrictions as a treatment for the disorder. They found that older children, already affected by PKU, improved on a diet that did not include phenylalanine. And by 1961 it was clear that if newborns with the disease avoided phenylalanine entirely, they developed normally. Here was a clear mandate for a neonatal testing program. Affected children could be cured by dietary control. By 1968, after

MOMENTS OF TRUTH IN GENETIC MEDICINE

passage of a series of state laws, most newborns in the United States were subjected to a simple blood test in the first few days after birth. Public health genetics was a powerful new reality because of PKU.

Promoters of human genetics of this period were convinced that physicians should be trained systematically in genetics. When McKusick organized the first Bar Harbor summer course, in 1960, his intent was to "teach the teachers." He proposed that a "great vacuum" existed in the medical school curriculum. Students were learning genetics in short courses taught by persons with essentially no training in genetics: medical school faculties, dental school faculties, and other assorted experts from various professional schools. McKusick's course could provide these teachers with the technical skills their students needed in the anticipated new era of genetic medicine (McKusick 1981).

By 1973, McKusick could proclaim it a "commonplace" that "as infections and nutritional diseases are better understood, controlled and managed, congenital and genetic disorders assume great relative significance" (McKusick 1973, xiii). A 1971 chart prepared by Hymie Gordon of the Department of Medical Genetics at the Mayo Clinic showed that overall infant mortality from diarrheal diseases dropped precipitously after 1945 (penicillin became available to the general public in the United States in April 1945). Fewer babies were dying of infectious disease. At the same time, the number of babies dying as a result of congenital malformations or genetic disease stayed the same; the numbers were almost stable from 1915 to 1965. This meant that these deaths constituted a larger proportion of the whole. As a percentage of all infant deaths, congenital malformations rose, from 7 percent in 1915 to 15 percent in 1965 (Gordon, in McKusick 1973). Gordon's chart captured the critical argument: genetic disease as a proportion of the total disease burden was rising in industrialized nations because many other forms of disease were coming under control.

In this interpretation, some health problems could literally become more heritable with time. Overall neonatal mortality fell in industrialized nations after 1945. Some babies still died, but they tended to be those who had genetic problems of some kind. One study in Italy found that interventions such as better prenatal care, antibiotics, and maternal education improved the survival chances for babies across the spectrum of birth weight and term length. Eventually, only the most genetically vulnerable died. Environmental modifications thus made heredity more important as a cause of the heterogeneous phenomenon "neonatal mortality." Genes began to cause a larger proportion of the deaths. As Barton Childs has proposed, "Heritability goes up and down" (Childs 1999, 227).

The sensitivity of heritability to historical change provided a contextual explanation for the new relevance of genetic disease after 1960. The degree to which any health problem was shaped by heredity depended on the environment. After 1945, and increasingly in the 1960s, the successes of public health and antibiotics had transformed human disease. Disease had become genetic as a result of techno-historical change.

One other event in the 1960s had a marked effect on the cultural fortunes of genetic disease. This was the rise in the mid 1960s of the use of amniocentesis for prenatal diagnosis of chromosomal disease, specifically, Down syndrome (Rothman 1986; Rapp 1999; Harris 1975; Milunsky 1973, 1979 and various years). This technical capability intersected with the abortion rights movement and the eventual U.S. Supreme Court decision in *Roe v. Wade* in 1973 to put prenatal genetic testing at the center of corporate and medical interest in genomic disease.

In the 1950s, amniocentesis was used as a diagnostic tool for determining the severity of fetal hemolytic disease. Procedures for third- and second-trimester amniocentesis in rhesus isoimmunization were reported in *Lancet* in 1952 (Bevis 1952). By the mid 1950s amniocentesis had been used to detect Barr bodies in fetal cells, a technique that permitted the identification of fetal sex. Two Canadian histologists, Murray Barr and E. G. Bertram, published a 1949 paper reporting that sex chromatin could be seen in many kinds of cells from human females. The sex chromatin turned out to be an inactivated X chromosome, present only in females. Four separate research groups immediately applied the technology to prenatal diagnosis (Cowan 1992). And just as rapidly, the diagnosis of fetal sex was used in families with a history of sex-linked disorders such as hemophilia. Prophylactic abortion of a male child, of course, meant accepting the 50 percent possibility that the male fetus was healthy (Riis and Fuchs 1960).

Human cytogenetic techniques were also undergoing a period of rapid change. Extra chromosomes or abnormal chromosomes had been linked to several diseases, and with improvements in laboratory methods, human karyotypes (the images of the complete complement of chromosomes from a single person) were much easier to interpret. Between 1965 and 1967, amniocentesis was used in the United States to predict about nine cases of chromosomal disease in fetuses. These diagnoses, which made possible the selective abortion of affected fetuses, received "a considerable amount of publicity," according to one participant at a 1967 conference on the "intrauterine patient." The methods of extracting amniotic fluid and growing and analyzing cells from this fluid were "not sufficiently reliable" to justify the high risk of miscarriage at the time, but the technique appealed to the medical and scientific community and to the public in general.[7]

Participants at that conference explicitly targeted abortion law as a barrier to the control of genetic disease. "Most of the considerations of this presentation will be of limited significance if more liberalized medical abortion laws, similar to those which have just been passed in England, are not instituted in the United States" (Klinger and Miller 1968, 81). Abortion was thus understood to be part of a linked technological system.

For some observers, the combination of amniocentesis, prenatal diagnosis, and selective abortion would make possible new, more acceptable and legitimate forms of eugenics. Population pressures, which attracted considerable attention after the publication in 1971 of Paul Ehrlich's *Population Bomb,* dictated that no couple should bear more than two children. In 1970, Bentley Glass, then president of the American Association for the Advancement of Science, proposed in a lecture to the association that the right to procreate was no longer paramount. "The right that must become paramount is the right to be born with a sound physical and mental constitution, based on a sound genotype. No parents will in that future time have a right to burden society with a malformed or mentally incompetent child" (quoted in Paul 1998, 137). A few years later, the New York University pediatrician Joseph Dancis echoed this perspective in an essay in a volume reflecting on the future of medical genetics. "In recent years the feeling has grown among both physicians and the general public that we must be concerned not simply with ensuring the birth of a baby, but of one who will not be a liability to society, to its parents, or to itself. The 'right to be born' is becoming qualified by another right: to have a reasonable chance of a happy and useful life. This shift in attitude is shown by, among other things, the widespread movement for the reform or even the abolition of abortion law" (Dancis 1973, 247). For Dancis, and perhaps for many others pondering genetic disease, the movement for legal abortion was not about women's rights. It was about the rights of the unborn, who should not be brought into the world with genetic burdens.

Between July 1969 and October 1973, the Prenatal Birth Defects Center at Johns Hopkins carried out 156 amniocenteses on pregnant women at risk for bearing children with serious genetic or chromosomal disorders. Like many other medical centers, Hopkins offered this service to high-risk patients who would qualify for medically necessary abortions, at a time when abortion on demand was illegal. Fifty-seven of these women already had a child with Down syndrome, forty-eight were at higher risk of bearing a child with Down syndrome because of their age, and eighteen had already borne a child with Tay-Sachs disease. The Hopkins center "explains to the couple the nature and risks of the disease, but never exerts pressure to terminate the pregnancy. Diagnosis is confirmed on the dead abortus for the reassurance of the parents rather than for scientific reasons."[8] The center charged couples $295 for the procedure, but this did not cover the cost of providing all the necessary expertise and support, including counseling and obstetric care. And so the difference was made up with funding from the March of Dimes and with donated time and services from physicians interested in the potential of amniocentesis.[9]

After *Roe v. Wade,* genetic diagnosis of the fetus became a medical industry. Aubrey Milunsky's slim 1973 book about the prenatal diagnosis of hereditary disorders grew through the 1980s and 1990s into a thick, multi-editioned guide to the medical management of "genetic disorders of the fetus" (Milunsky 1973). Most of these disorders were managed by selective abortion. Indeed, the selective abortion of affected fetuses is a practical, cost-effective, and reliable way to control genetic disease. It has been and remains the primary intervention of genomic medicine.

The fortunes of genetic disease in the past century, then, are tied to shifts in public policy, including changing abortion laws and state-mandated neonatal testing. They are linked to technological innovations such as improved methods of karyotyping and the development of amniocentesis. They are embedded in the rise of the biotechnology industry and shifts in patent law. They reflect new professional coalitions around particular diseases or disease categories, such as mental illness or metabolic disorders. And they have been shaped by antibiotics, which reduced the overall medical impact of infectious disease. Genetic disease is also intertwined with personal histories, family stories, community action groups, state record-keeping practices, and the rules and standards of biological registries. Technical knowledge of human heredity is a critical part of this process, but it does not move through time in isolation. It is imbricated in multiple domains.

Although one might more conventionally distinguish nineteenth-century ideas about a vague, morally loaded constitutional illness from twenty-first-century notions of genomic disease by invoking technical legitimacy, I want to distinguish them by their cultural productivity. Constitutional illness provided nineteenth-century physicians with an explanation that patients understood and accepted because it conformed to prevailing conceptions of the links between moral status and bodily integrity. Constitutional disease was situated within a narrative of moral improvement and self-restraint. Genetic disease after 1945 engaged not with moral suasion but with technological optimism. It provided scientists, health care professionals, and patients with a model for disease that reflected the expectation that technological intervention was both possible and appropriate. Genomic disease is bound up with commodified technology and visions of unrestrained biomedical intervention. Human cloning, if it does occur, is unlikely to become an important means of reproduction. But the imagined human clone is a predictable human product of guaranteed, certified quality. The clone embodies the promise of control that is central to genomic medicine. The clone exemplifies the perfect cure for genetic disease.

By the 1990s, the idea that all disease is genetic disease had acquired considerable public legitimacy in the United States and in other industrialized nations. This power was expressed in policy debates, popular culture, scientific funding, and elite technical writings. The expansive version of genetic disease being promoted included, literally, everything. Francis Collins echoed other genomics promoters when he proposed that all diseases are genetic, with the possible exception of "some forms of trauma."[10] Genes are now widely believed to play a critical role in cancer, heart disease, obesity, mental illness, and many other states of bodily and social pathology. Scientific papers tracking genes for intelligence, personality, musical talent, or sports ability are also common. It is in fact difficult to find any trait or behavior of any social visibility that has not been linked in scientific and popular sources to heredity, including divorce and television viewing habits (Nelkin and Lindee 1995).

Gene frequencies in a breeding human population, of course, do not measurably change over a forty-year period. Rates of obesity and divorce have changed quite a bit in the United States since 1960. But as I learned from an obesity researcher, it is still true that some people get divorced and others do not, and that some people are obese and others are not, and genetics is the study of variation in a population in whatever shared environment that population experiences. In the contemporary climate, those genetic factors that might be involved in any and all individual differences can loom large, regardless of the undeniable environmental influences.

This idea may have limited relevance to everyday clinical encounters. Physicians do not necessarily base their clinical care on the more enthusiastic claims of James Watson, Francis Collins, Lee Silver, or other promoters of genomic science. Many practicing physicians may have had relatively little training in genetics or may find that whatever training they did have (they were learning the polymerase chain reaction [PCR] techniques by the early 1990s) was of little relevance to their everyday diagnostic and caregiving duties. Patients by the 1990s were commonly asked about family histories of cancer or heart disease, and their answers might well have some effect on clinical care, leading perhaps to an additional test or accelerated testing schedule. But even the most enthusiastic champions of genomic medicine recognize that genetic cures are not yet an option for most conditions. Somatic gene therapy, in which corrected genes would be inserted into cells in malfunctioning organs, has proven to be much more difficult and dangerous than originally expected. One gene therapy research subject has died; several others acquired leukemia as a result of the therapy. Nor have the expected profits materialized. Many biotechnology firms produce research findings rather than consumer products, with their profits based on expectations and projections, not on delivered goods. A few dramatic therapies grounded in molecular research do not constitute a revolution in clinical care.[11] And even journalists have begun to notice that the Human Genome Project is not generating clinical miracles. In practice, therapeutic interventions based on genomic medicine remain extremely limited.[12]

Yet a revolution might still be coming. At the dawn of the twenty-first century, pharmacogenomics promises tailored drug treatments sensitive to the precise genetic qualities of individuals. DNA-chip technology promises screening options for fetuses and newborns that were unimaginable even a decade ago. Preimplantation diagnosis and selection promise a technological solution to the political and social problem of abortion. Many bioethicists seem to have joined the biotech revolution enthusiastically, proclaim-

MOMENTS OF TRUTH IN GENETIC MEDICINE

ing the inevitability of germ-line manipulation and human cloning, often on the grounds of consumer autonomy and the rights of patients to buy whatever medical services they desire. Indeed, consumer autonomy and choice drive Lee Silver's imagined future world, in which humans branch into two different species, one genetically engineered for excellence and the other merely human. Genomic medicine might not yet dominate the clinical encounter, but it does dominate public culture and biomedical research and plays an important role in mapping the human future.

In Donna Haraway's exploration of the dog genome and dog breeding, the dogs come close to domesticating the people. Haraway's wolf-dog hybrids are resourceful and powerful, drawing on human garbage for their own advantage and convincing "early man" to help out with the whelping (Haraway 2003). Like the wolf-dogs prowling the human settlement, afraid to come closer and weighing the trade-offs, industrialized populations with access to genomic technologies seem to be in the process of overcoming fears of cloning, germ-line gene therapy, and other biotechnologies. In the process, they begin the technological domestication of the human species. And in this case, as in so many others before it, domestication is not strictly a laboratory science. Genomic medicine has been brought into being by many social actors. It is not a linear descendant of either scientific genius or scientific hubris. It does not come from one position on the board and move down or up to another. It is instead a communal project, the logic of the pack.

The transformations during the very fertile period I examine here involved the emotional, intellectual, technological, and social labor of many actors. Hard work made transcendent genetic disease a powerful reality, and those doing this work included technicians, legislators, voters, psychiatrists, Amish farmers, physicians, geneticists, mental health professionals, and patients and family members. The knowledge they produced was of many different kinds, and it could and did sometimes move across epistemological boundaries. Emotional knowledge acquired by a parent in close, continuous contact with a child with a complicated genetic disease became a resource for physicians and geneticists. Religious records compiled for one purpose could be reconfigured as pedigrees. Common sense or folk knowledge could be used to justify technical enterprises. Social experience could be a surrogate for blood testing. Political pressures could enhance the importance of technical interventions. And underlying this communal project was an implicit shared expectation of technological solutions.

The community of technical practitioners interested in hereditary disease in the period I examine drew on new laboratory technologies that could help localize disease to specific sections of the genome. They expected understanding and control through biochemistry or cytogenetics. Genetic disease could be seen in the cell, detected in the bloodstream, and prevented or treated through genetic counseling or dietary controls. They also called on genetic disease to explain more complicated and important disease categories. The diseases that had seemed constitutional in the nineteenth century became biochemical products of underlying genes, which could in theory be identified and controlled. At the same time, disciplines such as behavior genetics and human cytogenetics attracted new followers and acquired the standard trappings of new academic specialties, such as societies, journals, and international meetings. Relatively relaxed human-subject rules permitted rare, obscure diseases to be tracked through large hospital data sets. Neonatal mortality from infectious disease continued to fall, bringing into clinical sight infants with rare genetic diseases who would not have survived without antibiotics. This facilitated the creation of that great catalogue *Mendelian Inheritance in Man*, which began in 1964 with a short published list and had evolved by the early twenty-first century into a website with more than thirteen thousand entries, and growing. If McKusick's catalogue is a proper meter, most genetic diseases now known were identified and named after 1965.

I begin in chapter 2 with a study of the establishment of the first neonatal testing program, for the genetic metabolic disorder phenylketonuria, as it developed in the late 1950s and 1960s. A restricted diet for the treatment of PKU was tested in the 1950s, and its success justified a massive public health program. But the urine test then used in Britain to detect the presence of the disease in newborns depended on a network of health workers who saw every baby at home at about six weeks of age. The practices of hospital delivery and health care in the United States required a test that could detect PKU within the first three days of life, before babies and their mothers were discharged from hospitals. When Robert Guthrie and his laboratory assistant Ada Susi developed a blood test that could detect PKU in a two-day-old infant, many state public health departments began testing newborns. As state legislators became interested and committed, they facilitated the creation of the basic framework of the heel prick, the thick filter paper, the state laboratory system, and the rules for interacting with par-

MOMENTS OF TRUTH IN GENETIC MEDICINE

ents. This system could and did accommodate new diseases. Neonatal testing varies from state to state but now commonly includes testing for sickle cell anemia, maple syrup urine disease, galactosemia, and congenital hyperthyroidism. In chapter 2, I explore what instructions were provided to technicians who took the blood samples, what rules states developed for managing the blood of newborns, what new mothers were told about the test, and how the U.S. Children's Bureau organized and assessed the rapidly developing PKU testing program. I focus on the many layers of work involved in making PKU, and by inference all genetic disease, a compelling public health problem.

In chapter 3, I look at provenance and pedigrees in the genetic field work of the 1960s. Biological anthropologists, medical geneticists, and physicians began carrying out increasingly technical field work with special human populations around the world. The Johns Hopkins University medical geneticist Victor McKusick, one of the most important figures in postwar human genetics, was a skilled practitioner of this field work, and I look at his early studies of Old Order Amish in Pennsylvania. In this relatively closed breeding population, he found high rates of hereditary dwarfing conditions. He also found among the Amish an intense interest in genetic disease and in disease more generally. The religious and secular records of the Amish facilitated his research, and McKusick's self-conscious use of the social to reach the biological is striking. His pedigrees were compilations of many kinds of stories. The pedigree remains an extremely important narrative frame in genomic science. No sample can be interpreted without a knowledge of its history and its familial embeddedness. I suggest that the stories people tell about their bodies and their families are a critical part of genetic science and that special, isolated human populations have sometimes been active participants in the scientific enterprise.

In chapter 4, I turn to a different kind of labor, the labor of standardization. My specific focus is the standardization of the human chromosomes at a time when they were very difficult to tell apart, before staining technologies developed in the 1970s made them appear distinctive. But my broader suggestion is that standardization in medical genetics attracted health care professionals to the emerging possibilities of genetic medicine. This is unquestionably true in relation to chromosomal disease. Physicians were obviously and openly attracted to the dramatic new images of human cytogenetics, after a method of preparing chromosomes described in a 1960 paper made them much easier to see and distinguish. The concomitant emergence of a morbid anatomy of the human karyotype also attracted medical attention. In the twenty-two months after Jerome Lejeune characterized Down syndrome as the consequence of an extra chromosome, at least twenty-three other chromosomal anomalies associated with disease were reported in the scientific literature. Cytogeneticists and physicians began looking at chromosomes as sites where disease was visible, transparent, unequivocal. Drawing on institutionalized populations, including people with disabilities or mental retardation and the imprisoned, they karyotyped almost eighty thousand people by 1975. Such a massive effort was bound to produce a more detailed picture of chromosomal disease. Chromosomal anomalies made sense of poorly defined clinical entities, and chromosomal similarities linked conditions hitherto believed to be distinct. This transdisciplinary and transinstitutional project required that every researcher looking at the chromosomes agreed about their identity, and I consider the processes involved in standardizing chromosomes, cytogeneticists, and human minds and bodies.

Chapter 5 tracks and engages with zygosity, technical expertise, social and personal knowledge, and the use of twin studies by geneticists in the 1960s. I focus on twins as a physical and social resource for human behavior genetics after about 1955. I am particularly interested in the creation of vast registries of twins around the world. These included Franz Kallmann's twins with schizophrenia, tracked through New York psychiatric institutes, and the National Academy of Sciences Veteran Twin Panel, developed after 1955 and eventually including almost sixteen thousand pairs of white male twins who had served in the armed forces in World War II. In 1961, a Swedish research group found that asking twins themselves whether they were identical or fraternal was as reliable as carrying out expensive blood tests. Twins, apparently, knew something about their own fetal origins. Twin research provides a compelling case study of the ways in which folk, social, and common knowledge are interwoven in scientific texts.

My last empirical chapter, chapter 6, focuses on the history of a disease that, despite its rarity and complexity, illuminates many of the forces and tensions propelling the rise of genetic disease to public prominence. The disease is familial dysautonomia (FD). Like many genetic diseases it is an autosomal recessive disorder, and it was recognized and named only in the postwar years. And like many genetic diseases, its interpretation, management, and technical assessment have been shaped by parental and familial involvement at many levels. In the spring of 2001, it became the focus of a patent fight. Like so much of the human genome, the FD gene has become a legally contested commodity. Familial dysautonomia is a classic genetic disease, fitting well within the narrowest possible definition. Its pattern of inheritance is understood; it appears in a genetically isolated population, Ashkenazi Jews; it is rare, strange, difficult to manage, and of limited importance in everyday clinical care. Most physicians will never see a baby with FD, but this is true of the vast majority of Mendelian disorders. At the same time, with the rise of genomic medicine, FD became something that could be owned, subject to patent law and contractual agreements about testing. The history of FD is virtually a microcosm of the overall history of genetic disease in the post-1945 period. It ends, as my larger story does, in the domain of ownership.

In my concluding chapter, I suggest that notions of genetic disease promised to solve or explain frustrating problems such as schizophrenia, to alleviate guilt for parents, to guide prevention through reproductive control, to lead to a cure for cancer, and to provide a technical model of disease that could explain differing patient outcomes. In a very real sense, at the dawn of the twenty-first century we live in bodies understood to be readouts of a master text that is a guide to personal health, success, talent, intelligence, and risk. This reality is expressed in people's self-descriptions, in legal decisions, in the institutions that market DNA technologies, in criminal justice investigations, in public support and funding for genomic science, in health care policy debates, and in scientific research papers in journals spanning the biomedical domain from cancer to AIDS. It is culturally and technically true, in every way that matters. It is the way we assess fetuses, discuss cloning, or explain our own life narratives. Perhaps other ways of thinking about the body, and other narratives that could be told about destiny, disease, or risk, are now difficult to imagine. But there have certainly been other narratives in the past, and it is safe to expect that there will be others in the future.

In 1966 Victor McKusick announced that genetic factors were involved in all diseases. He compared the mutant gene to a bacterium or virus and emphasized the importance of cataloguing the many forms of genetic disease (McKusick 1966 and various years). Almost forty years later, McKusick's proposal has indeed become the conventional wisdom. Many textbooks and survey articles express the idea that with the increasing control of infectious disease in the developed world, genetic diseases assume a greater importance in health care. Genes are "of growing importance in current clinical practice in the developed countries," and the same will be true in the developing world "once the high mortality rates due to infection and malnutrition come under control" (Weatherall 1991, 1). Or, "in the past, the major causes of diseases and death were infection and malnutrition, and the genetic causes did not attract much attention. As social conditions have improved and medicine has begun to control infections, the genetic causes of ill health have become much more important" (Nora et al. 1994, 4).

This way of seeing genetic disease is a product of many cultural choices. The physician Karl H. Muench, in his 1988 textbook on genetic medicine, proposed that genetic information was becoming more important because of research and new knowledge, but also because of "a social revolution with respect to sex, marriage, reproduction and the roles of the sexes" and because of the interest of the lay press, which was "publishing a profusion of information concerning these matters" (Muench 1988, 2). The emphasis on genetic disease that is so familiar in the wake of the Human Genome Project is not a strict mandate of biology but a cultural choice enacted on many levels. It is a decision to locate disease in technologically specified heredity and to emphasize the genetic components of all disease.

The rise of genetic disease to scientific and cultural prominence was, therefore, not solely an intellectual event. It had social, institutional, emotional, and ideological dimensions, and these fed back into the intellectual and technical work. The idea that genetic information could explain all human disease was reinforced and validated by the bureaucratic systems sustaining neonatal testing, by medical enthusiasm for cytogenetic analysis, by the massive databases of information on twins or abnormal chromosomes or isolated populations such as the Amish, and by the successful use of fetal testing to prevent the birth of children with genetic disease. Knowledge of genetic disease was produced in a feedback loop of politics, theory, laboratory technique, human suffering, and historical contingency.

The scientific conclusions were trustworthy in the end nonetheless. Genomic medicine produces reliable, usable knowledge, and although I am examining the processes involved in making scientific truth and considering the social and cultural networks that drive this process, my study is not an unveiling project that construes the embeddedness of technical knowledge

MOMENTS OF TRUTH IN GENETIC MEDICINE

as a form of pollution or contamination. I have no theory of contagious magic. I think folk knowledge can and does move into technical knowledge without undercutting the legitimacy of the technical. I know that this has happened often enough that accounts of such circulation appear routinely in elite scientific and medical texts. Authors commonly describe what they have learned from patients, local knowledge-producers, informants, public records, family members, or community leaders. Knowledge in biomedicine, and perhaps in other sciences, comes from many explicit sources, and more than once technical expertise has validated folk constructions that originally framed the technical problem. Both information and ideas move through the feedback loop with multiple starting points.

Yet what does this mean for our understanding of the knowledge-making process? Is there any meaningful sense in which biomedicine is made "from the bottom up"? And could recognizing the patchwork qualities of biomedical knowledge facilitate alternative ways of interpreting the body or interpreting disease? I think my work suggests some of the raw potential for change that this process incorporates. In consequence, this is a faintly optimistic book, proposing that the people participating in knowledge production are much more widely dispersed than is generally admitted, that knowledge percolates up and down through culture, and that some wholesale, even indiscriminate, borrowing hides in plain sight in certain standard narratives (the formal scientific text, for example). It proposes that this borrowing can be readily excavated with the mundane critical tools of the historian or social scientist and that such excavations can illuminate important aspects of the process of making knowledge.

It would be possible to argue that injustice is the direct global cause of more human disease than is heredity. Racism, economic exploitation, environmental degradation, and political oppression are implicated in the public health crises of the developing world. People may differ in the rate at which they starve to death, as a consequence of hereditary variation in metabolism. Does that make starvation a genetic disease? Sir John Sulston, former director of the Wellcome Trust Sanger Center in Cambridge, England, and prominent genomics researcher, has made a related point: "Will the benefits of genomics extend to everyone, or only to the rich?" Sulston notes that two billion people lack access to basic medical care, and only 10 percent of global research and development funding goes to the diseases that account for 90 percent of the worldwide disease burden.[13] Injustice is not the necessary province of molecular genetics, and there is no obvious connection between the existence of the Human Genome Project and of world hunger. But the World Health Organization and United Nations data explored by Sulston do at least suggest that genes do not cause all human disease. Hereditary disease is of growing importance in the total disease burden in privileged, well-fed, and well-cared-for populations living in powerful nations. Certainly genes do cause many terrible disorders and do seem to play a complicated role in differential risk for other conditions such a cardiovascular disease, cancer, or some forms of mental illness. But the story told here of increasing access to genes and increasing control of genetic disease is not solely about what causes most human disease. Like the human body more generally, genes are implicated in geopolitical organization and networks of industrialization and commodification. The modern rise of genetic disease to cultural prominence reflects something other than a strict quantitative calculus of the causes of human disease.

Finally, I must address directly the question of increased knowledge. One way to understand the new importance of genetic disease would be that, with our increased knowledge about heredity and DNA, the relevance of genetic disease has become clear. Armed with this new knowledge, this explanation would suggest, physicians and scientists can recognize the roles of genes even in extremely complex states of pathology. It is certainly true that scientific knowledge about human heredity has expanded greatly in recent decades. Biotechnology firms now vie with public research programs to map the genome, new genetic tests appear at a rapid pace, and new drug therapies are marketed to persons genetically at risk for the common afflictions of prosperity. Yet to construct this new knowledge as an explanation of itself is to turn the cart around the wrong way. We do not generally know things because we know them. We know things, particularly natural things, because we have sought to know them.

Much more is known about DNA, genes, and genetic disease than was known in 1945, but why do we have this particular knowledge? Why do we know these things and not other things? What sorts of institutional, intellectual, technological, economic, and political frames made this kind of knowledge important, profitable, desirable? What processes and what kinds of labor were involved in the emergence of new diseases or the reconstruction of older disease entities as genetic diseases? And how exactly was genetic disease negotiated and understood by physicians, geneticists, parents,

MOMENTS OF TRUTH IN GENETIC MEDICINE

patients, and elected officials? I would suggest that we know much more about genetic disease because of such negotiations and because of the shifts that facilitated the creation of that new knowledge or made it newly profitable. Exploring such questions can provide a way to think critically about biomedical science.

Genetics is the study of bodily similarity and difference, and these are intimate, highly charged matters. The science of human heredity addresses questions that play out with great intensity in everyday emotional life, and making genetic knowledge cannot be construed as solely a technical enterprise. My account focuses on the technical work but also keeps in sight those folk narratives of sameness, resemblance, and difference that continue to play a role in genetic counseling, clinical care, and family life.

I want to close by making explicit what may be in some ways obvious: hereditary disorders have no common qualities, except that they are hereditary. Genes cause diseases in many different bodily systems, and hereditary diseases are treated by many kinds of medical specialists, in oncology, pediatrics, physical therapy, nutrition, surgery, dermatology, dentistry, psychiatry, neurology, gastroenterology, and so on. There is no unique bodily system affected by genetic disease. It is a class of diseases that can affect any system, any organ, any aspect of human health. Similarly, genetic diseases range in severity from extremely mild conditions, barely rising to the level of inconvenience, to fatal diseases that kill all those affected. There is no signature for genetic diseases. They do not necessarily have anything in common, except that they are caused by genes. Yet this one common thread binds them tightly together, for it means that they can be predicted probabilistically or incontrovertibly (depending on the disorder) by the examination of DNA. They can be detected in advance of their bodily manifestations—sometimes far in advance, in the fertilized egg to be selected for implantation. And what is trivial or inconvenient can in theory be prevented as readily as what is fatal.

CHAPTER 2

[**BABIES' BLOOD**]

Phenylketonuria and the Rise of Public Health Genetics

One of the great success stories in the history of genetic disease was engineered not by geneticists, physicians, or biochemists but primarily by advocates for children with mental retardation. Over the ten years from 1953 to 1963, an obscure and extremely rare metabolic disorder became the focus of major public health programs in the United States and other industrialized nations. It emerged from the "medical ghetto" (Yoxen 1982) of genetic disease to become the center of a network of information, technology, social management, and legislation. Phenylketonuria (PKU), a disorder of phenylalanine metabolism present in about one in fifteen thousand neonates in the United States, attracted funding from national organizations devoted to children's health. It was the focus of dozens of scientific papers. A field trial of a new diagnostic test was sponsored by the U.S. Children's Bureau, and by September 1965 neonatal screening for PKU was established public health policy in thirty-two states. In twenty-five of these states, the screening was mandatory.[1] Thus, in a short period, a rare genetic disease came to be embedded in a dispersed system of medical and political management.

Explanations of why this happened have appeared in retrospectives by participants, in scientific reviews, and in scholarly studies by ethicists, political scientists, and historians (Paul and Edelson 1998; Scriver 1995; Faden, Holtzman, and Chwalow 1982; Bessman and Swazey 1971). Perhaps the simplest narrative portrays the rise of PKU screening as the consequence of new knowledge. Scientists began to understand how to intervene in PKU through dietary controls: infants with the biochemical disorder could avoid mental retardation if the amino acid phenylalanine (normally essential to development) were virtually eliminated from their diet. This new knowledge led naturally to a legislative and public health program. PKU screening was therefore the simple consequence of new technical and therapeutic capabilities (Scriver 1995; Centerwall and Centerwall 1965; Guthrie and Whitney 1964).

A more complicated story appeared in a 1975 National Academy of Sciences report that was critical of the new screening programs. In this report, PKU screening programs were the product of "fragmented, uneducated and hurried decision-making." State legislators were swayed by pleas from parents of children with mental retardation, and these parental appeals led lawmakers to approve public health programs with uncertain consequences, based on "unfounded claims" about the efficacy of a low-phenylalanine diet. "There was very little recognition of the implications for public policy, or for the impact on individuals who were screened, of the fact that PKU is a *genetic* disease" (Committee for the Study of Inborn Errors of Metabolism 1975, 92–93; emphasis in original).

From a rather different analytical perspective, the historian Diane Paul tells an ironic story of uncertainty, conflicting agendas, and ideological utility in the rise of PKU testing and dietary therapy. Many physicians were skeptical of the testing programs, which they interpreted as inappropriate state intervention into medicine. The dietary therapy was boring for the patient, expensive for the family, and difficult to manage and assess. And the screening tests were sometimes inaccurate. Yet PKU testing has been idealized, Paul notes, by those on both sides of the long-running nature-nurture debate. For critics emphasizing the power of the environment to determine human health, the dietary control of PKU is a prime exemplar of the idea that biology is not destiny. And for those promoting the power of genetic medicine and the importance of genetic disease, the success of PKU screening programs suggests the many benefits to be expected from genetic screening more generally (Paul 1999b).

These accounts explaining how PKU became a public health priority are compelling ways of situating a genetic disease that has had an unusually high historical and political profile. From one perspective, the disease became politically relevant because of technical knowledge. From another, political relevance was driven by responses to the problems of children with mental retardation and the social experiences of family members who testified before lawmakers. From still a third, it reflected complex ideological commitments to the power of genetic medicine or to the potential of environmental intervention.

Certainly there seem to have been many kinds of knowledge and many stakes shaping the rise of PKU testing. Controlling PKU required the expertise of dieticians, pediatricians, psychologists, experts in mental retardation, epidemiologists, biochemists, and technicians responsible for efficiently testing infants' blood. It also often involved the expertise of family members who had experience with PKU and with the difficulties of managing the diet. One group with relevant expertise, however, was almost wholly disengaged from these events. Human geneticists, who were at this very moment seeking to legitimate the relevance of their discipline to medicine, played a minimal role in the rise of PKU testing. PKU as a public health problem in practice belonged to the community built around mental retardation in children. It was a disease defined not by its mode of transmission (heredity) but by its subject population.

Testing for PKU has been widely promoted as a demonstration of the social and medical value of genetic screening. Once the testing system was in place, testing for other diseases was easier. Eventually even the problems associated with PKU testing were seen as evidence that more genetic testing should be done. Geneticists complained, for example, that the low-phenylalanine diet was never properly assessed in a controlled trial because it quickly became unethical to withhold such a promising intervention. The data on PKU were scattered and incomplete, and many critics saw the entire system of testing and treatment as founded on rumor, uncertainty, and anecdote. Geneticists therefore proposed that neonates should be routinely tested for many other genetic diseases for which there was no therapy so that the natural course and population frequencies of these diseases could be better understood before any therapeutic regimen was established as sacrosanct (Committee 1975). Thus the PKU program could justify an expansion of neonatal testing even by virtue of its flaws.

In this chapter, I explore the history of this high-profile, rare disease. I focus on the rise of PKU to public prominence and legislative attention. I also look at the material culture of the PKU testing system. These materials include a bacterial strain used in the inhibition assay for detecting PKU in newborns, *Bacillus subtilis* ATCC 6051 (the strain number designated in the American Type Culture Collection); the thick filter paper that absorbed the babies' blood; and the blood itself, which became a resource for public health genetics and a permanent collection of human DNA. Because I am interested in types of labor, I pay attention to the instructions provided to the technicians who took the blood samples, to pamphlets given to new mothers, and to the physical process of the management of newborns' blood. I propose that attending to such stories makes visible the many layers of work involved in making PKU and, by extension, genetic disease in general a compelling public health problem. And I suggest that intellectual, physical, and emotional labor and knowledge all were critical to the rapid rise of PKU to political prominence.

The Blood

Like many other genetic diseases, phenylketonuria came to medical attention as a result of parental anguish. Genetic diseases are intimate family matters. They are transmitted from parent to child, and parents are both advocates for their children and the cause (at some level) of their children's suffering. Some of the technical experts who played a role in scientific interpretations of PKU had intimate familial knowledge of mental retardation in children. Both its discoverer, Asbjørn Følling, and the developer of the widely used blood test, Robert Guthrie, had family members with the disease. The social experience of physical and familial connection with children having this disorder played a role in motivating technical explorations. Emotional knowledge, then, is implicated even in the fact that the disease came to be known and named and extracted from the mass of other forms of mental retardation.

Følling's extended family included several children with the disease, and Følling's work with PKU began after the mother of one of these children demanded an explanation for her child's mental retardation and peculiar odor (the "mousy" smell of phenylketonurics). A maternal concern with the close physical experience of the disease prodded Følling to a biochemical answer, and in 1934 he identified the disease as a familial disorder of phenylalanine metabolism. He found that the children's urine turned green when mixed with ferric chloride, as a consequence of the presence of phenylpyruvic acid in the urine (Følling 1934; Scriver 1995).

The British geneticist Lionel Penrose almost immediately suggested that a low-phenylalanine diet might be an effective treatment (Penrose 1934), and in 1939 the American biochemists George Jervis and Richard Block proposed the same idea (Paul 1999b). But the low-phenylalanine diet was not tried until 1951, when two British biochemists tested the diet on three small children, all of whom showed some improvement (L. Woolf and Vulliamy 1951). By 1955, two other groups of researchers had tried low-phenylalanine diets in affected children, with some reports of success (Armstrong and Tyler 1955; Horner and Streamer 1956). British physician Horst Bickel and his colleagues reported that a three-year-old girl they were treating stopped having convulsions. Bickel proposed that if the diet had been started earlier, mental retardation might have been avoided (Bickel, Gerrard, and Hickmans 1953; Bickel 1996).

The possibility of therapy and cure that this dietary intervention promised led almost immediately, in 1956, to a British screening program to test infants' urine for the presence of phenylpyruvic acid. This testing was facilitated by the British practice of home health visits. Phenistix, the urine test, was most reliable when used with infants six to eight weeks old, and in Britain infants were routinely seen by a health professional, in their homes, at about this age. But in the United States, most newborns left the hospital after two or three days, too early for an effective urine test, and home visits were not a part of American medical care. In the United States, the use of a different bodily fluid would better conform to existing medical practice.

Blood was to become the material sign of metabolic disease. It was the transferable, mobile, testable segment of a new body that could be readily integrated into the public health system. It was the currency of public health genetics, the commodity that health departments produced just as they produced information and clinical practices. When the microbiologist Robert Guthrie and his laboratory assistant Ada Susi developed a test that could detect excess phenylalanine in a newborn infant's blood, they provided the technological frame for a massive blood-collection program. There was an interaction between bodily fluids and health care delivery systems: a blood test became central to the American system not only because of Guthrie's ingenuity but because of the prevailing practices of infant care in the United

States. The test itself was modeled to fill an existing space, and then in occupying that space it made possible other kinds of tests that drew on the fluid collected. The Guthrie test helped create the legislative frame that redefined PKU and eventually public health genetics.

Guthrie, a microbiologist and professor of pediatrics at the University of Buffalo Medical School, had a son with mental retardation and a niece with PKU. In 1960 Guthrie began explicitly looking for a way to detect excess phenylalanine in infant blood. With the approval of William Welch, the director of the laboratory at the Newark State School in Newark, New York, Guthrie was able to use blood samples from this large population of children with mental retardation. Drawing on these samples, he and Susi developed a simplified agar diffusion test for the presence of phenylalanine.[2] The new blood test was an indirect test that calculated levels of phenylalanine based on bacterial growth under specialized conditions. A strain of *B. subtilis*, a ubiquitous soil bacterium that infected plants but not animals or humans, was mixed with an inhibitor that normally stopped its growth. To this mixture was added a small disk of filter paper soaked with blood collected from the newborn's heel. If the blood counteracted the inhibitor and the bacteria grew, the test was positive: the phenylalanine in the blood interfered with the ability of the inhibitor to stop bacterial growth, and the size of the halo of bacterial growth was taken as a quantitative sign of the level of phenylalanine.

Guthrie became a vocal promoter of the use of this test for neonatal screening for PKU. An outsider to the community of researchers interested in human metabolic disorders, he took his case directly to parents, legislators, and the press rather than appealing only to the biomedical community (Paul 1999b). Guthrie's authority thus derived partly from his social location as a member of a family affected by mental retardation, and Guthrie explicitly invoked this familial experience and this social network in his public appeals (Koch 1997).

Other experts also drew on familial knowledge by proxy. Richard J. Allen, a pediatrician at the University of Michigan, apparently did not even speak with his colleague at Ann Arbor, the prominent medical geneticist James V. Neel, before publishing a paper on PKU in the *American Journal of Public Health* (R. J. Allen 1960). At least, Neel was not on the long list of people Allen thanked in the paper, though he was one of the most well-known human geneticists in the United States. Allen did, however, mention in his

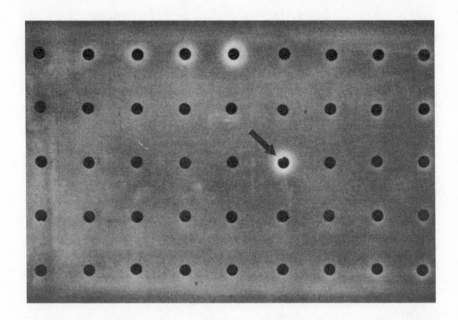

The bacterial growth halo in a positive PKU test, from Guthrie and Susi, 1963.

paper that he had spoken with a parent who had diagnosed his own child with PKU after listening to a lecture and acquiring ferric chloride–impregnated paper test strips (used in the urine test for PKU). And Allen did note that mailing ferric chloride kits to parents seemed to work well. "Lay persons can accurately interpret the color changes indicated by this method" (1664). Who had relevant knowledge of PKU? The eminent geneticist James Neel or a parent of a child with PKU? Who could be trusted to recognize its presence?

The Test

Guthrie published a short report on the inhibition assay for phenylketonuria in a letter to the editor of the *Journal of the American Medical Association* in 1961. His formal, peer-reviewed description of the test appeared in September 1963 in *Pediatrics,* though Guthrie had described the use of the inhibition assay in the control of PKU at a London conference in 1960 (Guthrie and Susi 1963; Guthrie and Tieckelmann 1962). The procedure was not inherently difficult, yet at every step it was possible to invalidate the

MOMENTS OF TRUTH IN GENETIC MEDICINE

test in some way. If the blood were taken too soon after birth, there would not have been enough time for phenylalanine to build up; ideally, blood would be taken when the infant was at least forty-eight hours old. Insufficient blood on the filter paper could make a second sample necessary, and getting that second sample involved the social labor of negotiating with the parents or perhaps contacting relevant physicians. Autoclaving the sample too long could destroy the phenylalanine. And in the interpretation of the semiquantitative results, skilled visual judgment was essential.

The heel prick for neonatal testing was and remains a demanding procedure. Infants can be squirmy, and infant blood is viscous and difficult to drop onto the Guthrie test card. In his instructions to technicians, Guthrie emphasized the importance of properly handling the process of taking blood, stressing that the blood should be applied immediately to the very absorbent filter paper. The spot of blood, after air drying, should be "at least 3/8 inch in diameter (but not more than 1/2 inch)," and it should be close enough to the edge of the paper that a disk could be easily punched out with a hole punch (Guthrie and Susi 1963). Later technicians and medical staff responsible for the test were provided with instructions for obtaining blood specimens that assumed that the parent would be present and involved in the test. The parent, in these instructions, was given the job of holding the infant, with the heel or toe lower than the body, while the technician chose an "area for puncture"—usually the heel—and pricked "deeply enough to secure free flow of blood." It was then necessary to "hold form BL50 [the Guthrie card] with printed circles uppermost until each circle is completely filled" with blood.[3]

The emphasis on soaking through the filter paper "so that when dried the circles of blood appear the same on back and front" reflected the persistent difficulties with this task. "Insufficient quantity of blood collected on filter papers" created a backlog of second tests in Maryland early in the program. Reliable analysis depended on filling the 3/8–inch circle entirely, and "repeated failure to do this" made a further sample collection and test necessary. "Investigation of the reasons for this problem shows that it occurs most frequently where personnel in the newborn nursery are rotated or changed."[4] Even at this first step, at this point of extraction by a standardized procedure, the workplace culture (rotation of personnel) interacted with the intractability of the newborn body (its tendency to be uncooperative, squirmy, difficult) and the quality of the blood itself (extremely viscous) to shape the utility of the test. Knowledge of genetic disease was also knowledge of how to hold a crying infant and puncture its skin.

The U.S. Children's Bureau of the Department of Health, Education and Welfare managed the initial field trials of the Guthrie test in the United States. As the federal sponsor of these trials, it had an interest in the reliability of the test in practice, not just in its scientific legitimacy but in its performance record in different laboratories and under different conditions. One of the bureau's earliest surveys found that labs varied in their skill levels and that the test seemed to be particularly prone to false negatives, a type of error that could undermine its value as a screening tool.

In March 1963 the Children's Bureau Central Laboratory prepared a hundred blood-impregnated filter paper specimens identified only by serial number and containing twenty-five samples with a blood phenylalanine level of 6 mg% or higher (positive specimens) and seventy-five with a level of less than 6 mg% (negative specimens). Forty laboratories that had taken part in the first field trials were asked to participate in this test, and thirty-six of them did so. Technicians were asked to record the values obtained for the hundred prepared specimens using the inhibition assay technique in the same manner routinely used to test the samples received from hospitals. The results, which Children's Bureau staff member Gladys Krueger emphasized revealed the precision rather than validity of the test, suggested that some labs were very good at properly identifying negatives and others were very good at identifying positives. In the final scoring, 60 percent (22 of 36 laboratories) got all or almost all of the positive samples correct. The remaining fourteen laboratories identified correctly anywhere from twenty-three to eleven of the positive specimens. Conversely, only fifteen labs correctly identified all the negative samples. There was a slightly greater tendency to misidentify the negatives than the positives.[5]

The Guthrie test made PKU a disorder that could be completely present in the newborn infant. Formerly diagnosable in the toddler through a suite of symptoms or in the eight-week-old with a urine test, with the Guthrie test the disease could potentially exist in the body of the newborn as early as forty-eight hours after birth. The timing of diagnosis of any disease is always important, but in the case of genetic disease timing is particularly dramatic. The technologies used to identify, for example, Huntington disease in a fetus change the meaning of the disease for everyone involved. PKU was an early example of a genetic disease that could be made present—in this case not

MOMENTS OF TRUTH IN GENETIC MEDICINE

through DNA testing but through blood testing—before it was a clinical syndrome.

This test also led, indirectly, to a definition of PKU as a racial disease. It was not that those who advocated testing believed that African American babies could not have PKU, but rather they believed that including African American babies in the initial field trials would make the test appear less cost-effective as a public health screening tool. The need to legitimate the testing technology provoked a particular way of thinking about PKU.

A Racial Disease

Penrose had earlier suggested that the disease had a racial component, though with a very different intent. In his 1946 paper in *Lancet* he emphasized the apparent absence of PKU in Jews, in an effort to make an antiracist point about group susceptibility to disease. When Guthrie proposed the field trial to the Children's Bureau on 2 May 1962, he simply stated, "It is suggested also that nonwhite infants should not be included in this study." This was one of the limitations on the study. Other limitations included a suggestion that the trial take place only in hospitals where "blood specimens can be collected as late as the third or fourth day of life."[6] A few weeks later the Children's Bureau sent a letter to states invited to participate in the field trial. Among the "points [that] should be read carefully" was an item 7: "The plan provides that the infant population used for this initial trial should consist of white infants who are three or four days of age (because phenylketonuria is virtually non-existent in non-white groups and because of our lack of knowledge as to what phenylalanine levels are in the first four days). Hospitals selected, therefore, should be chiefly those who serve white infants and who discharge such newborn infants only after three or four days."[7]

Two sorts of ignorance shaped this field trial: ignorance about the disease's racial distribution and ignorance about its manifestation in the first seventy-two hours of life. But the category "nonwhite" was more ambiguous than the category "infants at least three or four days old," as one regional medical director pointed out. Lucille J. Marsh, a physician in charge of the Children's Bureau office in Chicago, questioned the decision to exclude nonwhite infants. "This seems to me a mistake," she said, "even though cases are rarely found in non-white infants according to our present knowledge." It would nonetheless be valuable to have "some information as to the screening results among non-white infants." She also suggested that there was "presently so much confusion as to the percentage of white or non-white blood in many population groups as, for example, in the large Negro population, the Porto [sic] Ricans, the Mexicans, etc., that the division into white and non-white will probably be purely arbitrary in the next generation."[8]

Hormuth responded with the argument that nonwhite infants would skew the sample. The bureau was concerned about "getting an adequate sample of white infants where the expected incidence is approximately 1 in 20,000." There was only one reported case of PKU in a nonwhite infant, he noted, and if nonwhites were included in the study, he feared that "over half of the infants [sampled] might be non white and this . . . could lead to the mistaken conclusion that this assay is not effective as a screening tool." The exclusion of nonwhites was necessary in order "to give the assay a fair trial with a group where at least we have a vague idea of incidence." He acknowledged that there might be some cases of PKU among the nonwhite population but said that so far "existing screening and diagnostic methods have not turned up many cases." There were also some suggestions that nonwhite infants might have elevated blood phenylalanine without having PKU. The Children's Bureau would be "interested in exploring the question" but only as a separate project, not as a part of the initial screening trial of four hundred thousand infants to assess the Guthrie test. If "some states are interested in exploring phenylalanine levels in non-white infants (in addition to doing their share of white infants for the field trial)," this might be possible to arrange with Guthrie, who distributed the test kits. It would be important, however, wrote Hormuth, to "keep these figures separately. Urine screening of negro infants in well baby clinics thus far has not been productive."[9]

In 1963 the Children's Bureau policy attracted the attention of the federal authorities. The commission had received a complaint from Springfield, Massachusetts, "stating that non-white children were not being included in this program." A Mrs. Martin of the U.S. Commission on Civil Rights telephoned to ask whether the PKU program was sponsored by the Children's Bureau and the government was contributing funding. The staff member who spoke with her "explained that it was a screening program for the detection of PKU participated in on a voluntary basis by a number of state health departments" with funding from the Maternal and Child Health Program. Hormuth sent Martin a copy of the proposal in which Guthrie suggested that nonwhite infants be excluded. The efficiency or effectiveness of the Guthrie test could not be properly assessed with a group that was heavily weighted with infants who "as far as we know do not have this condition."[10]

By excluding nonwhite infants, Guthrie and the Children's Bureau staff were constructing their screening trial around a population they believed to be at higher risk for PKU. They were defining categories—"white" versus "nonwhite" babies—in ways that reflected their institutional concerns about justifying the Guthrie test as a mass screening technology.[11]

My point is not that the exclusion of "nonwhite infants" was malicious. In the context of this field trial, the decision to construct PKU in racial terms made sense. But the case does demonstrate how testing technologies can provoke particular ways of thinking about a disease. Marsh was right to point out that rates of PKU in nonwhite populations were unknown, but under these conditions, a "fair trial" of the testing technology depended on an interpretation of PKU in racial terms. The institutional and political priorities were bundled into the testing technology, foreclosing some questions while apparently answering others. PKU as a white disease drew on a readily applicable social category, widely recognized and extremely important to the organization of health care in the United States.

By late 1963 it seemed clear that PKU testing would be widely adopted by the states. This meant that the testing technology developed by Guthrie would be potentially profitable, and a pharmaceutical firm, Miles Laboratories, applied to the U.S. Public Health Service for the exclusive right to manufacture the test kits for states interested in PKU testing programs. The proposal was immediately nixed by the head of the Children's Bureau, on the grounds that states should be able to manufacture the kits themselves, at cost, as Massachusetts was already doing, and that the test had been developed with almost $500,000 in public funds from various entities and should therefore be readily available at a low cost.[12]

The Guthrie test had made it possible for PKU to become a public health problem, and this test, nurtured with public funds in so many ways, should therefore remain in the public domain. The profits to be made from PKU testing were the diffuse profits to society, not specific profits to corporate sponsors. The savings proposed by the identification of every baby with PKU were calculated in terms of the avoided costs of institutionalization and the enhanced contributions of wage-earning adults with PKU. Every baby with PKU was worth $100,000 in taxpayers' money, according to one 1962 estimate, which calculated both the cost of institutionalization ($2000 a year) and the "loss to the community in terms of productive capacity of these persons" ($5000 per year). PKU screening redeemed not only the child but "society," which was saved from an onerous burden of care.[13]

The Diet

The possibility of dietary intervention was the single most important factor encouraging legislative and public interest in PKU testing. That a special diet could prevent mental retardation in PKU was the one detail consistently included in materials given to new mothers and consistently invoked by legislators. It was the only biological fact, in these contexts, that mattered. Mothers were not told that PKU was a disorder of phenylalanine metabolism, that it was inherited as an autosomal recessive, or that it could be detected by an inhibition assay in which excess phenylalanine permitted bacterial growth. But they were told that it could be treated with a "special diet."

A Minnesota brochure, for example, emphasized that PKU damage could be prevented ("fortunately doctors can put these babies on a special diet"), and another brochure, from Findlay, Ohio, said that advocates for children with mental retardation "know that the best way of doing something about mental retardation is to prevent it from happening." This brochure had a title in lurid lettering asking new mothers, "How Will Your Child's Mind Develop?" The text itself emphasized the efficacy of the controlled diet.[14] And a Massachusetts leaflet told parents that "research workers have discovered that this mental retardation may be prevented if a special modified milk diet is started during the first few weeks of life . . . it is very important to find it early so that the special diet can be started before it is too late to be effective."[15]

The omission of details about the disease may have reflected the idea that the possibility of dietary intervention was all that mattered to parents and children. Mothers needed to know that effective intervention was possible, but they did not particularly need to know about the workings of the Guthrie test or the biological details of the disease. The brochures they received presented the disease in ways that could and did explain and justify the removal of blood from their babies' heels. This transgression of the bodily integrity of a newborn required a structure of explanation that made it benevolent.

The idea that the diet was a simple solution to a complicated problem was widely promoted in both official and popular sources. Such sources also suggested that PKU was a major cause of mental retardation in the United

States. Diane Paul quotes a 1957 *New York Times* article that proposed that "much" mental retardation was caused by PKU and another extremely rare hereditary disease, galactosemia. She also notes that in a 1965 congressional discussion of PKU testing, one senator proposed that "many" of the 5.5 million people institutionalized in the United States because of mental retardation had PKU, though the Children's Bureau had identified only 399 children with PKU over a four-year period (Paul 1999b, 9–10). Clearly PKU was not the cause of most forms of mental retardation or mental illness, but the successful dietary management of PKU did suggest that perhaps dietary control could be used effectively to treat other mental conditions. PKU was an example that aroused hopes in the mental retardation and mental illness communities. It was a genetic disease manageable through what seemed to be simple environmental intervention.

In practice, of course, the diet was not simple. It was not clear, when the PKU testing programs began, how long the diet needed to be maintained. Some physicians and scientists asserted that the diet could be stopped at about age six, when primary brain growth was completed (Hsia 1960, 1656). Others suspected that the diet would be a lifelong necessity, or at least necessary in women through their reproductive years. Nor was it clear that the diet would completely eliminate all symptoms of PKU, which included characteristic personality traits and skin problems. Furthermore, day-to-day management of the low-phenylalanine diet was complicated, expensive, and tedious for the phenylketonuric and the family. Most of those born with PKU and treated through dietary controls depended on a protein-rich commercially available formula, Lofenalac, that had been approved by the U.S. Food and Drug Administration in 1958. But this formula was not tasty, and although young infants seemed to tolerate the taste and odor well, older children sometimes resisted it. For parents, this resistance was itself a serious challenge. A 1977 Department of Health, Education and Welfare report on dietary management of PKU attended in equal measure to the quantification of food intake and the social training that parents would need to make the diet work And even by 1977, the age at which the diet could be terminated had not been determined. "There is much difference in practice in this regard" (Acosta 1977, 11).

Part of the uncertainty was a consequence of how promising the dietary intervention was. The diet had never been seriously assessed, because it quickly became unethical to withhold it. A few apparently successful cases—

fewer than twenty—fueled the PKU system. Data were incomplete, tentative, and unclear, and there seemed to be no obvious way of assessing the finer points of the therapy. Questions persisted about "effectiveness in relation to age or intelligence of the child at time treatment is initiated; the optimal length the treatment should be continued; or reasons for the success or failure of the child and his family in adjusting to the diet."[16]

In 1961 Gladys Krueger of the Children's Bureau staff proposed that those working in clinical facilities for children with mental retardation were rapidly gaining considerable experience with the treatment of children with PKU. Nurses, physicians, family support staff, and families were learning things about PKU by day-to-day experience and observation rather than through scientific analysis. But perhaps their fragmentary social and quotidian experiences could be amalgamated by the Children's Bureau in productive ways. The knowledge acquired through daily practice by persons untrained in statistical analysis or genetics could supplement the scientific data. If the bureau could work with a large number of clinics, Krueger proposed, it could combine their results in a comprehensive survey.[17]

A 1962 questionnaire to ninety-one special clinical facilities for children with mental retardation revealed 399 children with PKU admitted since 1957 (over the previous five years). Of these only 221 were under dietary supervision, and two-thirds of these had been admitted in 1960 and 1961. Children admitted earlier were far less likely to be treated with the low-phenylalanine diet: 1960 was the critical temporal boundary. There were also geographical quirks in the distribution of cases of PKU: three of the ninety-one clinics surveyed were responsible for 52 percent of the cases. Clinics in Ann Arbor, Los Angeles, and Philadelphia clearly made special efforts to find affected children and bring them under care. Staff members at these clinics reported that some families discontinued the treatment because they moved out of the service area, or found the diet difficult to manage, or decided on their own that the child was old enough to stop. Staff also said that treatment begun after age three had no effect.[18]

Based on this preliminary survey, the Children's Bureau decided that the population of treated patients with PKU was large enough to provide reasonable data on the effectiveness of the diet. But when the bureau sought expert advice on exactly how these data should be collected, it found considerable disagreement. In 1962 and early 1963 preliminary plans were distributed to a group of U.S. Children's Bureau advisers, and their responses

MOMENTS OF TRUTH IN GENETIC MEDICINE

included criticisms of the intelligence tests and profound concerns about the dispersed nature of the study, which would involve data from many different clinics and different people.

In general, the fourteen survey respondents said the Children's Bureau needed to collect more information about patterns of mental deficit, about the health of parents, about serum phenylalanine levels that could be unhealthy because too low, and about the difficulties for children and their parents in maintaining the diet. The information to be collected, then, would include data on the social characteristics of the child; a pediatric evaluation, including medical history, physical examination, and conditions other than PKU; biochemical test results; evaluation of dietary management; assessment of intelligence; and appraisal of physical and behavioral development.[19] The form for the initial visit recorded race; parental education, occupation, and ethnic descent; mental status of the child; urine and blood tests; record of developmental milestones (sat up, walked, spoke); and any symptoms of PKU, including odor, vomiting, eczema, and seizures.

This very detailed and time-consuming form also included places to record EEG (electroencephalogram) results, bone age (relative maturity of a child's skeletal growth, determined with x-rays), IQ, social responsiveness, attention span, language ability, muscle coordination, irritability, and what specialized products the child consumed to control phenylalanine intake.[20] A second form for follow-up visits recorded the family's "understanding of condition and diet plan," "ability to cope with actual diet provisions," "cooperation in carrying out diet plan," and "emotional maturity in handling problems with diet and with family members."[21] This form also recorded the results for the child of periodic intelligence tests, tests for signs of PKU, blood levels, EEG results, and a space for explaining termination of treatment if it occurred.

Even in this study, conducted in 1962, it was impossible to include a control group. When some respondents proposed that a group of children with PKU should not be placed on the diet, Krueger responded that although controls were essential to any study designed to evaluate the effectiveness of a particular treatment, controls in this study would be unethical. The value of the phenylalanine-free diet was already so well known—"the literature records extensive clinical experience substantiating that dietary treatment of children with phenylketonuria is beneficial to some degree in a significant number of cases"—that withholding such treatment would be "impossible to defend from an ethical point of view." All that could be done in this regard, Krueger noted, was to ask participating clinics to keep track of those patients who for one reason or another terminated treatment themselves. Family members and patients who voluntarily refused the diet would thus become crucial informants. They could violate the technical imperative of the PKU program and, in the process, provide valuable data otherwise unavailable.[22]

A further complication was that the natural course of PKU was not known. The disease was rare and had often been lumped together with all other forms of mental retardation. It had been described in 1934 but not studied systematically. Some clinicians familiar with the disease were impressed by its variability. There were occasional reports of children with PKU who made significant gains without dietary treatment. This did not mean that plans for dietary treatment should be abandoned, but a knowledge of such variability, said one pediatrician, would "make observations on the treated group of greater significance." Jack H. Rubinstein, of the Hamilton County Diagnostic Clinic for the Mentally Retarded in Cincinnati, noted that it would be "a great temptation to say or imply that whatever gains there are in the area of mental or physical growth and development in treated children are the result of their treatment." Based on his own observations, he was not sure that this would be valid. And June M. Dobbs, a physician at St. Christopher's Hospital for Children in Philadelphia, made a similar point: "I think that not only children who are on the low phenylalanine diet should be included in the study, but also children who are found to have phenylketonuria and are not on treatment. We are implying that improvement in the child's mental age is the result of the low phenylalanine diet. However, it has come to our attention that there are several children with relatively high IQs, that is, in the 60 to 75 region, with phenylketonuria who have never been on a low phenylalanine diet." Such children showed an increase in their mental age "as they are followed," she said, and it was therefore important to "guard against stating categorically that the diet alone has caused the child's level of functioning to improve. It might have been that this was a child with a relatively high IQ in spite of the diet."[23] Some clinicians who had a working familiarity with PKU thus questioned the diet's importance.

Phenylketonuria, then, in this comparative study, was a multilayered social, psychological, ethical, and even economic phenomenon. The family's ability to manage the diet was an integral part of the disease, as were various biological markers (bone age, EEG), developmental markers (sat, walked), and interior psychological states (irritability). The types of psychological tests chosen mattered to the assessment of the diet, as did race, ethnicity, IQ, and maturity. PKU was not just a matter of biochemistry. Like many other genetic diseases, it was an extraordinarily complex psychosocial experience of illness that required a network of support, both technological and psychological. The beautiful simplicity of the diet—the appealing idea that a genetic disease could be treated effectively with a single dietary restriction—contrasted with the day-to-day reality of living with such a profound bodily difference.[24] PKU was understood to be a disease that engaged an entire family's "ability to cope," and families were understood to be crucial sources of information about the disease. The diet's success was measured not strictly in terms of blood phenylalanine but in terms of a thick cluster of feelings, bodily function, and money.

The Children's Bureau Collaborative Project did not begin until 1967. By January 1974 it had collected 14,256 items of information, which were stored in a central data bank. These many facts, quantified, suggested that the diet was effective in attaining near normal IQs in children with PKU and that it also ensured "physical growth within the normal range of the American population" (Committee 1975, 31–32).

The purpose of all the Children's Bureau efforts was to assess the value of mass screening for PKU testing. But as the mass screening programs developed they moved beyond the Children's Bureau's control. Legislation on PKU testing was driven by anxious constituents, casual assessments by legislators, and popular interest.

Legislative Action

The primary organization driving the rise of PKU screening legislation was a voluntary health group consisting of parents and family members of children with mental retardation. The National Association for Retarded Children (NARC) was created in 1950 as a coalition of smaller, local organizations. Generally these groups sought to improve educational and therapeutic services for individuals with mental retardation and to encourage public funding for mental retardation programs. NARC exemplified the health-related consumer groups that were of growing importance to public policy and even to scientific research after 1945 and particularly in the 1970s and later. Its members had personal knowledge of the effects of mental retardation, and they sought to affect public policy through a variety of strategies including public information, official endorsement of particular programs, lobbying, and so on. Although some members were physicians, these often were members because they had experienced mental retardation in their own families (Committee 1975, 44–45). And NARC was firmly in favor of PKU legislation.

It is important to remember that the Guthrie blood test and the low-phenylalanine diet did not dictate any particular legal or administrative actions. Testing could have been left to individual physicians. This was the option favored by such physicians' groups as the American Academy of Pediatrics and the American Medical Association. Or testing could have been managed by departments of public health, as a measure reflecting new technical knowledge of public benefit. Legislation was not, strictly speaking, necessary for the benefits of PKU testing to be applied. But state legislators, encouraged by local members of NARC and also perhaps by voters who read about PKU testing, took up the campaign and quickly approved neonatal screening programs that would have profound consequences. The transparent value of the PKU diagnosis justified a blood-collection network that opened doors almost wholly unnoticed by those promoting legislation.

By 1974, forty-four states had PKU testing laws. Many of these were passed in 1965, which seems to have been the critical year. Such laws varied somewhat, but most made the testing mandatory, with religious objection by parents as the only basis for avoiding screening. And most were open-ended about what could be tested for: under Alaska's law, like some others, the state could test for any metabolic conditions that threatened health or intellect, a rather broad mandate. None of these statutes mentioned genetics (Committee 1975, 44–55).

I want to turn now to an examination of the way the state law was passed in Maryland, which approved PKU legislation in 1965. The initiative for PKU screening in Maryland came from more than one direction, and it reflected popular and political concerns as much as technical insights. Essentially, neonatal screening could not have happened without this broad interest and support, which expressed the social meanings of mental retardation and public faith in technical solutions.

Legislation in Maryland was not supported by the medical community, primarily because physicians (in Maryland and elsewhere) thought screen-

MOMENTS OF TRUTH IN GENETIC MEDICINE

ing should be voluntary. As one medical committee put it, "all of us strive for good medical care for our patients. To be ordered to do so by a state law is an insult to our cumulative medical intelligence."[25] But legislation was pushed by advocates in the mental retardation community who came to believe that a mandatory screening program was necessary, particularly after voluntary compliance proved to be low (Paul 1999b).

Statewide screening for PKU began in Maryland in 1961 with urine tests by local health departments, targeting children three years old and younger who attended the state's Child Health Clinics. In 1962, Maryland was invited, with other states, to participate in the U.S. Children's Bureau trial of the Guthrie blood test for neonatal screening. But Maryland health officials hesitated. "Perhaps you have wondered about Maryland's reluctance to enter the program which many states have begun involving the use of the Guthrie test for phenylketonuria in newborn infants. We hope that the enclosed statistical analysis of Maryland hospitals will serve as an explanation of this reluctance." The statistical table showed that in only five hospitals in the state was the stay of full-term infants four days or longer, and the total number of births in these hospitals represented less than 10 percent of the full-term births among the white population. "On the basis of this analysis, we feel that the yield from this program in Maryland would hardly justify the cost involved."[26]

Instead, Maryland Health Department officials were trying to "educate our practicing physicians regarding the importance of urine testing for PKU from the age of six weeks until, at least, the age of three years." Urine screening in Maryland (independent of the Children's Bureau program) had, incidentally, picked up one case of PKU in "a six month old negro infant in one of the Baltimore City Child Health Clinics."[27]

Later the next year, Maryland's attitude toward the Children's Bureau screening program changed. Benjamin White, chief of the Maryland Department of Health division of community services for people with mental retardation, said the state would test twenty thousand infants as part of the Children's Bureau program by 1 July 1964.[28] Around the same time, State Senator J. Albert Roney, Jr., took an interest in PKU testing and told a Maryland health commissioner that he had heard about it from a "client whose wife is expecting a child." The constituent asked the senator whether the state had considered passing a law to require a test on newborn babies, "which he advised me was known as the 'PKU Test.'" Roney had heard that it was possible "to take corrective measures to alleviate, correct and prevent further mental illness, and correct such illness if apparent, as a result of this test." Should Maryland explore such testing?[29]

Health Commissioner Perry Prather quickly replied that the Department of Health was indeed conducting PKU testing at four hospitals and was hoping to expand the program to all Maryland hospitals. But he added that "in regard to legislation, we believe that little is to be accomplished by making this test a requirement by law. Instead, we hope to make its importance known to the practicing physicians in the hospitals of the State."[30]

Deputy Health Commissioner Edward Davens later elaborated on this matter in reply to another interested legislator. He noted that in 1963, both NARC and the American Academy of Pediatrics passed resolutions on PKU, and both statements "strongly recommend against legislation which would name specific disease entities or especially specific laboratory tests." The rapid evolution of scientific testing methods, he said, made such specificity inappropriate. "Both resolutions do back legislation which would call for swift application of newly emerging screening procedures after they are scientifically acceptable by providing the necessary laboratory capabilities and other things necessary to apply this knowledge to all our newborn infants."[31]

In late December 1964, representatives of the Maryland chapters of various medical groups, including the Academy of Pediatrics, Academy of General Practice, the Obstetrical and Gynecological Society of Maryland, the Maryland Association for Retarded Children, the Maryland Hospital Council, and the Maryland State Department of Health, met to discuss PKU screening.[32] The general consensus among these representatives was that voluntary screening would be "more effective" than compulsory screening.[33] "It was agreed that in Maryland the goal of screening newborns could be accomplished most effectively by voluntary action rather than by legislation."[34]

In none of these documents do advocates of voluntary testing spell out what would be more effective about voluntary screening, and at least some of those appealed to were skeptical. One physician noted that "unless the groups responsible for accreditation and approval of hospitals include the testing as one of the elements" needed for approval, "the voluntary compliance will be spotty." He pointed out that "one of the greatest forces will of course be medical-legal, but this requires the presence of defective children damaged by failure to use the test, and I do not consider that to be a satisfactory substitute."[35]

MOMENTS OF TRUTH IN GENETIC MEDICINE

By this time two legislators wanted to introduce a bill for mandatory screening, regardless of what the health professionals were saying,[36] and on 8 April 1965, Governor Tawes signed House Bill 895, which made screening for PKU in Maryland compulsory. The text was notably brief, less than a full page, and it specified that whoever filed the certificate of birth was also required to "cause to have administered to every such new born child a test for phenylketonuria." It was, in other words, a disease-specific law. It included a sentence permitting parents to refuse the test on religious grounds. By December 1965, the Maryland Department of Health, formerly opposed to the mandated screening, began to produce a *PKU Newsletter* for health professionals, and the December issue proclaimed, "Thirty-two states with laws for PKU screening: Guthrie test being utilized by over 2,000 hospitals."[37]

The initiative for PKU screening in Maryland thus came from more than one direction. It came from a federal bureau of the Department of Health, Education and Welfare, from health professionals within state offices who valued the test but were reluctant to support legislation, and from casual conversations between constituents who had pregnant wives and their legislators (an index of the visibility of the disease in popular culture). So, too, did interpretations of PKU screening reflect different perspectives. Legislators were drawn to the image of effective, rapid state intervention and politically mediated medical intervention; physicians saw screening legislation as an intrusion; health department officials worried about how it would work in Maryland's hospitals.

Meanwhile, just as the legislative mandates began to accumulate—with more and more states passing PKU legislation and accepting the idea of mandatory state screening of blood drawn from the bodies of newborn babies—the U.S. Children's Bureau began to have doubts about the enterprise it had sponsored and promoted.

Public Doubts

By the late 1960s, significant problems had emerged with the PKU program, including ambiguous test results, problems with maintenance of the diet, confusion about the disease itself, and doubts about the effect of diet on intelligence and psychosocial development. The PKU program itself made these problems visible. More children were tested, more were placed on the diet, and more were assessed by mental health professionals as they matured. The disease that had blended into the background, only one of many to be found in populations of children and adults with mental retardation, now emerged into sharp focus, and like most diseases it proved to be more complicated than it had once seemed.

"A critical and objective review of the experience of the past fifteen years now indicates that the problem may not be as simple as was originally thought," wrote John A. Anderson and Kenneth F. Swaiman in a brief note introducing conference proceedings in 1966. The "spectrum of the basic and clinical aspects of the disease is wide." The disease itself was a more diverse entity than originally believed. Some children with high phenylalanine levels were being "confused with children who have classic phenylketonuria." In addition, some children who did seem to have "classic phenylketonuria" (whatever that might be) developed normally even without a restrictive diet. "The result of more thorough studies on family constellations suggest[s] that the number of these mild or untreated cases may be a far greater segment of the clinical spectrum of the disease." The time had come, the authors concluded (in 1966), to "redefine not only the basic variations in the primary enzymatic defect, but also . . . to clarify the significance of the biochemical screening procedures now so widely employed." Diagnostic criteria were in disarray, the sharp dichotomy of heterozygote/homozygote was "doubtful at the chemical level," and some patients had high phenylalanine for a different reason and therefore did not have PKU but a "new condition" (Jervis 1966). Furthermore "'occult' phenylketonurics" had been discovered in the British screening program—patients who had PKU but whose excretion of phenylpyruvic acid in urine was too low to give a positive reaction in the Phenistix test (L. I. Woolf 1966). And at least one set of identical twins, both with a PKU genotype, had been studied in which one twin was of normal intelligence and the other severely retarded. "The difference in clinical manifestation cannot be genetic in origin if [these twins] are monozygotic" (580).

If the new claims about PKU and its variations were true, then perhaps some of those being treated with a low-phenylalanine diet did not in fact need the diet. And for a person who did not need the diet, the consequences were unknown. One observer suggested that unnecessary treatment seemed "a small price to pay for preventing the mental deterioration otherwise inevitable" in other patients. But Samuel P. Bessman, professor of pediatrics at the University of Maryland Medical School, was becoming a vocal critic of PKU testing programs and proposed that this was "another mistake" (L. I. Woolf 1966, 58; Bessman 1968).

More disturbingly, it was also becoming clear even by 1963 that the test was picking up genetically "healthy" infants who did not "have" the genetic disease PKU but were born with mental retardation because of maternal PKU. A paper in the December 1963 issue of the *New England Journal of Medicine* described three women whose high blood phenylalanine levels apparently produced mental retardation in all of the fourteen children born to them (Mabry et al. 1963). None of these children had genetic PKU, but their bodily condition was produced by the same biochemical processes. What disease exactly did these babies suffer from? And who was being tested when the newborn's blood was taken? "This is the first published evidence that phenylketonuria in the mother may be a significant cause of mental retardation in the offspring, due to the damaging effect to the foetus before birth caused by the mother's high phenylalanine blood level," Guthrie wrote in an internal memo in 1964.[38]

Phenylketonuria, the disease itself, as an intellectual and clinical category, was being called into question as a consequence of the details produced by a widespread system of disease management. What exactly was PKU? How could it be identified? Perhaps the screening program was defining the bodily condition inappropriately, perhaps even in ways that could damage children. The three-day Conference on Phenylketonuria and Allied Metabolic Diseases in Washington, D.C., in April 1966 included papers on primary metabolic questions, dietary management, and the role of government and legislation in the management of health problems. PKU was defined in these different contexts as a enzymatic defect affecting the central nervous system, a public health problem, a consequence of patients' refusal of an unpalatable diet, a psychological problem, and a test of screening technologies and methods that could guide future screening programs for other diseases.

The Geneticists

In 1970, after PKU screening programs were virtually universal in the United States and in many European nations, the geneticists began to get involved. Some of the leading figures in postwar genetics, including James V. Neel at the University of Michigan, James F. Crow at the University of Wisconsin, and Joshua Lederberg at Stanford University, took an interest in phenylketonuria. A committee of the American Society of Human Genetics appointed to study the problem also included Alexander Bearn of the Cornell Medical Center, Barton Childs of the Department of Pediatrics at Johns Hopkins University, and Robert F. Murray at the Howard University College of Medicine.[39]

Many of these professionals with a specific stake in genetic disease were skeptical about PKU testing programs. They wondered whether legislative intervention in a medical matter of this kind was appropriate. They thought neonatal screening programs varied too much from state to state. They wanted to bring reason and order to public discussions of genetic disease. And they wondered whether the blood collected for PKU screening could become a resource for human population genetics or for the study of rare genetic diseases. If I wish to claim that PKU testing played a role in the rise of genetic disease to medical prominence, I must at least acknowledge that the testing program did not look particularly promising at first to physicians or even to geneticists. PKU testing posed vexing problems to those who knew the most about hereditary metabolic disorders and who had the most direct stake in increasing public interest in genetic disease.[40]

Their concerns reflected at least partly a predictable discomfort with PKU testing becoming institutionalized without much input from geneticists. But they also recognized that this program would be the model for other programs and, indeed, that other genetic testing programs were already underway even by 1971. A whole field of bodily surveillance was growing up almost spontaneously, in ways that reflected individual interests and regional variation as much as technical knowledge and rational planning.[41] The geneticists decided to address the problem directly and to provide, if they could, the leadership that neonatal testing as a systematic endeavor seemed to need.

The Study of Inborn Errors of Metabolism—SIEM, as it came to be identified—was formally detailed in a 13 December 1971 proposal by the National Research Council to the National Science Foundation. The study's authors promised to "develop an effective program for dealing with inborn errors of metabolism as a single identifiable but multifaceted health problem of national importance."[42] The hodge-podge historical development of neonatal testing was the key problem: innovations had been put into practice and programs instituted for early detection of some conditions without any reasonable broad perspective that could make sense of the programs. "Characteristically these programs were established to deal with a single defect."

There were thus "separate fund-raising activities, screening programs (some required by state statutes), public information projects, research programs, etc., for phenylketonuria (PKU), sickle cell anemia, cystic fibrosis and others." While such efforts were clearly of social value, they were, viewed from the national level, "not without serious faults."[43]

In effect the proposal was criticizing the organization of science, policy, law, and medical practice around particular disease entities—the process I have been examining here. Instead, the proposal stated, organization should be structured in terms of a larger category, namely all metabolic defects that had been defined by 1972 (there were 165). Individual diseases that had historically framed medical research were not the proper units of analysis, this proposal suggested, partly because so few people were affected by any one genetic disease. Breaking down metabolic disorders into individual disease entities tended to mask their medical importance. "Taken individually, as they usually are today, inborn errors of metabolism are rare. But valid statistics are lacking on total numbers. By conservative estimate, 30,000 afflicted infants are born annually and as additional defects are defined this number will increase. Each year 6 percent of all hospital admissions are due to diseases caused by the defects which were not detected early enough for institution of treatment."[44]

In the committee's 1975 book describing their research, the preface made clear some of the professional and disciplinary tensions involved in the reframing of genetic disease. The text proposed that public health professionals had inappropriately taken over testing for genetic disease: "Screening programs for genetic diseases and characteristics have multiplied rapidly in the past decade and many have been begun without prior testing and evaluation, and not always for reasons of health alone." The committee predicted that many other screening tests could be expected in the future as disease patterns changed, preventive medicine became a high priority, and new genetic research provided more information about hereditary disease. The "mistakes already known to have been made" combined with these conditions to suggest a need for the committee's study. The goal was to "minimize the shortcomings and maximize the effectiveness of future genetic screening programs" (Committee 1975, p. iii).[45]

By the 1970s, other genetic diseases had become the focus of national and regional legislative initiatives. The U.S. Congress passed the Sickle Cell Anemia Control Act in 1972, in a move that reflected both the racial politics of the period and the increasing "celebrity" status of sickle cell disease (Wailoo 2001, 165–96). The passage of this bill in turn provoked representatives of other ethnically specific genetic diseases, such as Cooley's anemia (a type of thalassemia), to clamor for equal recognition.

Conclusion

The blood extracted from almost every newborn in the United States for more than forty years is a physical monument to the rise of genetic disease to medical prominence. Often stored in state laboratories and sometimes made available to private industry for various forms of blood testing, the babies' blood is an archive in which a particular construction of the body is preserved. The blood itself is a record, not only in the sense that it records data about individuals but also in the sense that its systematic management makes manifest some central properties of the biomedical construction of the body. Fragmented for the purposes of obtaining information, the body moves into an information-retrieval system and becomes a form of transportable data. The blood itself is also a sign of faith that, through testing, redemption can be achieved—the sick can be healed, the mind restored. The child with PKU can be prevented from experiencing the degraded state of mental retardation that is as a consequence of the blood itself, which makes state intervention possible by providing the information needed to justify action.

Genetic disease acquires particular properties not only through political debates or medical assessments but also through the intimate management of blood. The management of babies' blood is a laboratory replication of the social management of PKU. It is a controlled, choreographed reenactment of a larger political drama, and it has its own uncertainties and problems (measuring the bacterial growth halo; getting the blood to soak entirely through the filter paper; handling the Guthrie card; teaching new lab personnel the methods) and its own economy (should the test kits be privately licensed or publicly produced? Should tests be "wasted" on "nonwhite" infants?). PKU is not, in my account, solely an internal bodily state; it is also something manifest in culture, texts, lab results, and legislation.

As problems emerged with the technological parameters that shaped PKU testing and diet, issues that seemed to have been resolved became contentious. Yet the institutions grounded in these technologies persisted and became the framework for other genetic tests. *Certainty* about both test and therapy gradually faded, but the legislative frame built from that certainty persisted unaffected and could be adapted to other uses. A poorly understood and poorly defined disease entity with highly variable expression and requiring a difficult treatment therapy became, in the 1960s and 1970s, a great victory of applied human genetics. Geneticists were minimally involved in the early history of PKU testing, but their discipline benefited from this success even as their academic societies produced a report critical of neonatal screening programs. A genetic disease with plasticity, amenable to environmental control, presented the ethical prospect of effective intervention with such force that questioning the test or the intervention became unethical. Some uncertainties and questions about PKU disappeared because they were no longer asked or no longer open to being asked. Other uncertainties emerged as difficulties with the performance of the Guthrie test and the management of the low-phenylalanine diet became clear.

How, then, was PKU made into a major public health problem? Certainly relevant were technologies that conformed to the health care delivery system. The Guthrie test suited the standard practices of maternal and infant health care in the United States. Also relevant were the legislative decisions that focused on a particular clinical interaction. Many new medical interventions become possible and legitimate without ever becoming the focus of state-by-state legislation. PKU was unusual in that it was politically relevant and important long before it was a well-understood clinical condition. And it was the possibility of redeeming children affected by PKU and saving public funding that drove this legislation. A simple blood test was all that was needed to transform a few children's lives. The actual savings, given the low number of PKU births, were almost irrelevant. But they justified state legislation.

For PKU, there was no sharp line between the bodily experience of disease and the systems that managed and made sense of it. It is not possible to extract PKU from the technologies that reveal it. The technician who autoclaves a Guthrie card is participating in making PKU, as are the laws that mandate tests for PKU and the manufacturers who produce Lofenalac.

The historical moment I have explored here is a local one: the pricked heel, the babies' cry, the full red circles soaking through the filter paper. I consider the disease as a physical manifestation in the bodies of newborns, as an elevated level of phenylalanine, detectable by a semiquantitative inhibition assay, present in a blood sample obtained at least seventy-two hours after protein intake, and as a social and legislative phenomenon manifest in draft resolutions by medical groups, pamphlets distributed to new mothers, new state laws, lab manuals for technicians working with the Guthrie test, and follow-up survey proposals. I have followed both the blood and the texts, asking how PKU was defined by both narratives and technologies.

More generally, I have explored some of the processes through which genetic disease acquired the properties it possesses today as both proximate and ultimate cause of all disease and as a public health problem on a grand scale. The PKU case helps to make visible the awkward, piecemeal nature of this transformation. It is a success story, unquestionably, but also a story of negotiation and decision making in the absence of a complete understanding of the disease in question. PKU screening is the best-case scenario. It is a serious genetic disease that can be detected early so that those affected can be spared the consequences of their biochemical differences. It has had unquestionable benefits for those born with this metabolic disorder. Even those critical of PKU programs from different perspectives recognize that screening produces a net social good. Yet threaded through this success story are all the complexities and ambiguities that characterize genetic screening more generally. Genetic diseases commonly have variable expression, so standardized screening tests may not be enough to understand a particular case; treatments can be difficult and unpleasant even if they work; a truly effective program can create undue public enthusiasm, facilitating expanded testing for conditions that cannot be treated; and like most technologies, genetic testing and therapeutic intervention can have unintended consequences, producing, for example, a higher rate of PKU in the overall population as people with PKU do not have mental retardation and thus are more likely to marry and reproduce (Cowan 1992).

By 2004, all U.S. states supported mandatory screening for PKU, congenital hypothyroidism, and galactosemia, and, all but Idaho, screening for sickle cell disease. Thirty-two states screened for maple syrup urine disease; twenty-eight for homocystinuria. Many states also screened for congenital adrenal hyperplasia and glucose-6–phosphate dehydrogenase deficiency, a sex-linked genetic disease. A few had added HIV and toxoplasmosis to the screening test. Transforming these testing programs was a new technology, tandem mass spectrometry, which permits screening for up to twenty disorders in only two minutes, using a single drop of blood. A newborn in the United States is thus immediately subject to a wide range of tests, some of which identify disorders for which no effective treatment is known.[46]

The data produced by the Human Genome Project can be expected to continue to expand the range of genetic testing. A steady stream of new disease genes can quickly be accessed as textual signs of risk and future pathology. The ability to test asymptomatic children for late-onset disease has been the focus of particularly intense concern among ethicists and health professionals. And proposals for DNA-chip technology promise a comprehensive genomic screen, at birth, through which many different kinds of bodily states can be recognized or predicted. Still unclear are the benefits of knowing, from birth, that one has an elevated risk for some form of cancer in the fourth or fifth decade of life. As the PKU case suggests, technologies close options as well as open them and prevent inquiry as surely as they encourage it. As we embrace genetic testing and a model of all diseases as genetic—as genetic disease emerges as the central public health problem of the twenty-first century—it is perhaps useful to pause and consider how the bodily and social meanings of genetic disease reflect available networks of technology and social management.

CHAPTER 3

{ PROVENANCE AND THE PEDIGREE }

Victor McKusick's Field Work with the Pennsylvania Amish

Provenance is the record of the "ultimate derivation and passage of an item through its various owners."[1] It is commonly used to describe the history or pedigree of a painting—who has owned it, its value at various stages of ownership—but it also has a meaning in silviculture, in which it refers explicitly to genetic stock. Provenance, for forestry professionals, is the record of where a seed was collected and the character of the "mother trees." In this chapter, I explore provenance in both senses, as a record of ownership and as a record of genetic stock. The ownership to which I refer is both biological and intellectual. Someone's body yielded every blood sample collected in the vast postwar project of human genetic population research. Every sample belonged to a certain person, and that person's identity was specified in some form in the textual record built around the blood sample. Someone had originally "owned" the blood when it flowed in their veins, and though the blood changed hands and moved to a laboratory, it retained an identity linked to the original owner. But ownership is also a way of characterizing the experience of knowing something: just as blood came from specific persons, so too did data, evidence, and interpretation. And like blood, these bits of knowledge were sometimes linked textually to their original source, trackable through the archival and even the published record to the persons who had originally experienced them or had originally come to know that they were true.

My primary focus here is on Victor McKusick's field work with the Pennsylvania Amish in the early 1960s, especially his work with a rare form of dwarfism, Ellis–van Creveld syndrome. But McKusick's field work was only a small part of a much larger project in postwar human genetics. In the 1950s and 1960s, human geneticists and other technical experts carried out relatively large-scale studies of human populations around the globe. Geneticists from the United States and from the scientific centers of Europe generally traveled to stay briefly (for a few weeks or months) with an isolated population of some kind, and while in the field they collected blood and other biological materials, including urine, feces, breast milk, tissue samples, hair samples, and teeth. They also collected family narratives, asking those they studied to describe their family trees, with special attention to cases of disease. They took photographs or made drawings, and they often measured the people they were studying (height, weight, arm and leg length, torso length). Then they returned with the drawings, photos, notes, and tissue samples to their home laboratories, to assess the data and prepare publications.

Such practices of field collection and data analysis had begun in biomedical research as early as the 1910s. The Brazilian geneticist Francisco Salzano reported in 1957 that there were already ninety-five published scientific papers focused on blood analysis of South American indigenous groups, dating back to the 1930s. Geneticists and other scientists had tracked down isolated groups, convinced individuals to submit to a blood test, labeled the blood to be sent to a laboratory for processing, and aggregated the findings to reach conclusions about human history. The Carib, Guajiro, Piaroa, Guahibo, Arawak, and Caramanta had all been bled and tested—439 Pijao Indians in Colombia were subjected to blood testing for a paper published in 1944, and almost three thousand Andean Indians of various groups in Ecuador for a paper in 1952; also studied were the Quechua of Peru, Tucano of Brazil, Alkuyana of Surinam, Matacos of Argentina, Maca of Paraguay, and Panzaleos of Ecuador. These populations had thereby been physically brought into technical explorations of race, migration, mutation, and "white admixture." This sort of research, then, was not new in 1950. But it does seem to have accelerated after 1950. This acceleration might have been facilitated by enhanced technological capabilities at two unrelated levels: laboratory analysis of blood grew more informative, and air travel improved.

What were geneticists expecting these remote, isolated, or "primitive" populations to reveal? Some geneticists, such as James Neel, were looking for clues to human evolutionary processes (Lindee 2003). Others were interested in geographical migrations and historical relationships between racial and ethnic groups. And others were interested in genetic disease. They tracked visible anomalies, such as Ellis–van Creveld syndrome in the Pennsylvania Amish or albinism in the Hopi of Arizona. And they tracked geographical anomalies, such as the presence along the Pacific Rim of small populations that appeared to be African. Identifying suitable populations, assessing their genetic status, learning their reproductive histories, and extracting from them blood, tissue samples, and family histories—all were important activities in postwar human genetics. Reproductive isolation made recessive diseases visible; religious interest in genealogies facilitated the construction of pedigrees; and geographical concentration simplified field research. Some populations made themselves into scientific resources by virtue of their practices of marriage, family structure, and religious faith.

The Amish were among the most productive of such populations. Since 1960, the Amish have been the focus of hundreds of genetics research publications, and results derived from these studies have been referred to in many thousands more.[2] Geneticists found new genetic diseases in the Amish, including forms of dwarfism, neurological disorders, albinism, and malformation syndromes. They also found high rates of previously characterized, rare genetic diseases such as Ellis–van Creveld syndrome, phenylketonuria, Weill-Marchesani syndrome, and Prader-Willi syndrome (McKusick 1978). Later the Amish were the focus of a sustained and ultimately unsuccessful search for the gene or genes responsible for manic-depressive illness (Gerhard et al. 1994; Ginns et al. 1996; LaBuda et al. 1996).

The Amish have been useful research subjects as a consequence of their own social choices and religious values. They are the descendants of Swiss Anabaptists who emigrated in 1690 to the Rhineland in southwestern Germany and then from 1714 to 1770 to the United States, in response to religious persecution (McKusick 2000, 203). In some geographical locations, the founding population was extremely small. Lancaster County Amish in Pennsylvania, for example, are believed to be descended from a group of

MOMENTS OF TRUTH IN GENETIC MEDICINE

no more than two hundred people. Studies of genealogy, gene frequency, and family names suggest that the Amish groups in Ohio, Indiana, and Pennsylvania do not constitute a single breeding population. Although they share practices and beliefs, they rarely intermarry (McKusick 1978, 10–11).

Those practices and beliefs are themselves constitutive of genetic isolation. As the world changed around them after their immigration into the United States in the eighteenth century, the Amish rejected some new technologies such as electricity, the telephone, and the internal combustion engine. They observed simplicity of dress and lifestyle. And they observed strict rules about marriage and social interaction. In 1960 there were about forty-three thousand Amish in the United States and Canada and none in Europe. Whatever permitted the Amish to survive as a distinctive cultural group was apparently unique to North America (Hostetler 1963).

They married within the group, and their children were educated not at public schools but in Amish schoolrooms. They maintained a keen interest in family lineage. In 1978, McKusick found sixty publications by the Amish detailing family records and lines of descent. These included the massive 1957 Fisher genealogy, a book providing data on thirty-six hundred families in Lancaster County (McKusick 1978; Fisher Family 1957). "The interest of many simple peoples in genealogy is a matter of note," said McKusick in his assessment of this corpus of family histories. The Bible set the precedent, and the Amish genealogies were a form of religious expression, much like the extensive Mormon genealogies (McKusick 1978, 12).

Geneticists, like simple peoples, were also interested in genealogy. McKusick's field research in the early 1960s with the Old Order Amish in Lancaster County tracked a rare form of hereditary disease, Ellis–van Creveld syndrome (EVC), a dwarfing condition. Within a few years he had identified as many cases of the syndrome in the Pennsylvania Amish alone as had previously been reported in the entire medical literature (104). As he continued to work with the Amish he found many other recessive conditions, and he and his colleagues published many papers on genetic disease in Amish populations. But I focus here on his early work with the Lancaster County Amish and his efforts to understand one rare genetic disease. Although EVC in the Amish often appears in medical genetics textbooks as a demonstration of the founder effect, I emphasize this work not because it is somehow pivotal in itself to the history of genetic disease. McKusick was a pivotal figure, but his work with EVC would probably not rank in any standard historical account as his most significant contribution to biomedical knowledge. Yet this field project provides a window into the many kinds of labor involved in the construction of the legitimate human pedigree. And it is the pedigree as technical text, social document, and historical record that is at the center of my interest.

I suggest here that the pedigree seamlessly blends folk, emotional, social, and technical knowledge, compacting multiple perspectives into a single image and text. And I note that the pedigree is the bedrock tool of human molecular genetics. A geneticist cannot assess a sample of DNA without some knowledge of its history or provenance, its embeddedness, where it was found or acquired, and how the body it was from is related to other bodies, as brother, cousin, attacker, or victim. Stories of provenance, origins stories, are therefore inside the molecular laboratory. The pedigree is itself a form of provenance: it tracks the origins of a gene that causes a genetic disease through generations, providing a dense record of the gene's movement and previous ownership. It also contains evidence of who knew that the gene had appeared, who saw it, who recognized the disease. The provenance is intellectual and biological, thought and matter, words and blood.

In his field work in the early 1960s, McKusick drew on field methods from anthropology, sociology, and history. His labor and the labor of his informants turned the Amish into a medical and a scientific resource. Gossip, x-rays, feelings, blood tests, and social consensus were resources for the construction of the pedigree. Any data point might have more than one axis running through it, from notes in Bibles to state public health records to reports from the local undertaker. Knowledge of diseased heredity was craft knowledge, dependent on diverse diagnostic and social skills, documents, and practices.

McKusick and his co-workers used a wide range of ragtag materials to effect a retrospective reconstruction of events inaccessible through more direct means. Like the historian confronting a checkered archival record, he pulled together the available data to tell a particular, highly focused story. This story took the form of a human pedigree, an inscription thickly layered with contingent details. Here is the crucial issue: the moments of truth in the pedigree were experienced by many different people, and they reflected many forms of record keeping and remembering. McKusick's careful, self-conscious use of the social and the cultural provided access to the biological. The sheer mass of specificity, depth, and detail overlapping in each data point was the functional equivalent of technical analysis. The pedigree is thus a striking representation of the mutually constituted categories of nature and culture. By attending to the field work that builds the pedigree, I suggest something of how that mutual constitution works. EVC, the disease, produced dramatic, quantitative bodily signs (dwarfism, extra fingers) that were visible because of the social networks that accounted for people, and these social networks, in turn, made possible the construction of a picture of underlying genetic causes. They made possible the capture of the biological through the exploitation of the social.

The human pedigree had long been a suspicious document in the eyes of some geneticists, with various calls over the years that human genetics should find a way to go "beyond" it. The pedigree was burdened by its transparently social nature, its dependence on the words of the subjects describing their parents or grandparents, and perhaps even by its connection to the project of eugenics and to Charles Davenport's questionable data-collection practices.[3] In the biomolecular era, and with the explosion of new work in human population genetics, physical anthropology, human cytogenetics, cancer genetics, and related fields, the pedigree was being remade into a resource for laboratory science.

As Yoshio Nukaga and Alberto Cambrosio (1997) point out in their ethnographic study of the pedigree in contemporary genetic counseling, pedigrees "still constitute the basic investigative tool" in human genetics. The stories people tell about their families move from a "web of oral narratives to a sequence of visual inscriptions which, in turn, become part of larger inscriptions connecting medical pedigrees to the visual display of, say, cytogenetic or molecular biological test results" (32). Even the most technical, machine-driven inscriptions of molecular genetics are grounded in the social complexity of the pedigree, which is nature-culture and represents a signal case of the employment of cultural resources to achieve the erasure of contingency.

McKusick's papers from this period contain hand-sketched maps and directions, telling field-workers where particular houses were and where families lived. At the same time, field-workers were mapping relationships between families, on similarly scribbled pieces of paper also tucked away in archived files. Both forms of maps, depicting roads and landmarks (general stores, silos, and red barns, across many miles) and depicting complex familial lines (darkened and clear circles and squares, across generations), were ways of organizing the Amish.[4]

In Edward R. Tufte's explorations of envisioning information, he distinguishes between pictures of nouns and pictures of verbs. Maps and aerial displays, he says, "consist of a great many nouns lying on the ground," whereas pictures of verbs involve "the representation of mechanism and motion, of process and dynamics, of cause and effect." Extrapolating from Tufte's construction, the directional maps of Lancaster County are "nouns lying on the ground" and McKusick's beautifully messy and tentative genealogy maps, hand-sketched and tucked in folders and notebooks as the work progressed, are arguments about cause and effect and stories about history and heredity. They depict a temporal flow chart that makes its case using the "smallest effective difference" between diseased and not diseased, male and female, alive and dead (Tufte 1997, 73–78, 121–27). They are powerful, dense records of labor that engaged not only McKusick and his confederates at the Johns Hopkins University and beyond but the entire community. Both the Amish and McKusick were proficient collectors of genealogical data and trackers of disease. They co-constructed Amish genetic disease, as McKusick's voluminous and rich archival records make clear. In the process, they also participated in the construction of a mandate for medical genetics that has had profound consequences.

Victor McKusick and Population Studies

One of the remarkable things about McKusick's life story is that he has so effectively managed it that many of the same details and ideas appear and reappear in profiles of him, regardless of the author. These include the suggestion that his interest in human genetics derived from his being born an identical twin; the idea that hard work as a youngster on his family's farm in Maine contributed to his ability to work with the Amish; and the apologetic description of his work with cardiovascular sound, which was a scientific "dead end." McKusick is the author of many compelling life stories (scientific, professional, personal), and his skills as a storyteller challenge my own historical skills as I have struggled to both accept his own narratives and attend to those elements of the events in question that are subsumed or obscured by those narratives.

McKusick's medical training and status as a clinician shaped his approach to medical genetics.[5] Born in Maine in 1921, an identical twin, he remembers a childhood of books and farm chores. An abscessed arm during his freshman year of high school resulted in ten weeks of treatment at a

MOMENTS OF TRUTH IN GENETIC MEDICINE

Boston hospital, an experience that led him to become interested in medicine. The coat colors of the cattle on the family farm were the subject of his first genetics paper, written while he was an undergraduate; this was submitted to the *Journal of Heredity* to no effect.[6]

When he arrived at Hopkins as a medical student in early 1943 he was already interested in genetics and in history.[7] A military obligation led him to a two-year stint with the Public Health Service in cardiology and shifted his attention for about a decade to the sounds produced by the heart. He began using a spectral phonocardiograph to record heart sounds, and he wrote a textbook, *Cardiovascular Sound in Health and Disease*, published in 1958. He was also looking at the heart in sickle cell anemia and in Marfan syndrome (McKusick 1954) and collecting data about connective tissue disorders in general, a project that led to the publication, in 1956, of his *Heritable Disorders of Connective Tissue* (now in its fifth edition, 1993).

Organs and systems were leading him to a study of hereditary disease (or he was choosing organs that could do so), and his professional profile, by 1951 or 1952, was clearly that of someone with expertise in the relatively marginal field of human genetics. By 1957 he was the head of what was then the leading program in clinical genetics in the United States, the Division of Medical Genetics at the Hopkins Medical School. In 1959, in an act of some prescience, he organized a medical symposium, with sponsorship by IBM, focusing on the analysis of human genetic linkage with the assistance of digital computers.[8]

Around the same time, he sought funding for a study of a geographical isolate on Tangier Island, located at the southern end of the Chesapeake Bay and economically linked to Baltimore, where, at Johns Hopkins Hospital, "seriously ill patients and patients with puzzling illness" came for diagnosis.[9] McKusick had visited Tangier Island and spoken with the inhabitants. He knew the postmaster and some of the social practices ("On Saturday evening everyone dresses in his best clothes and parades up and down the narrow streets of the settlement. Enormous volumes of Coca Cola are drunk").[10] The inhabitants of the island were poorly educated and had limited diets, but they had access to medical care ("The island has had a physician continuously for over a century"); there were "some instances of striking longevity," based on the dates on tombstones.

McKusick proposed that the research workers in the study should rent a large house "built by a Captain Thomas of the mailboat family" on the island for the four months of the study, in the summer and early fall of 1958. All nine hundred inhabitants of the island would be examined, in the course of ninety working days, an average of ten examinations each day. "The program will be presented as a medical survey—which it is." School teachers would be hired to line up the people for tests and studies. These tests would include blood typing and other blood tests. The examination would also include a complete tape-recorded medical history, physical examination, dietary history, photographs of the naked subject in both color and black and white (the color photos so that hair and eye color were recorded), acquisition of a specimen of hair, a note on handedness, and a "complete pedigree." "Simultaneous with the studies of individuals, what records are available in the parsonage and at the Accomack County Court House will be searched. Information on gravestones will be copied."[11]

In this early proposal, five years before McKusick was to begin his work with the Amish, the basic structure of his later field work is evident. He proposed to collect from a special population details about family history and about bodily state, using general medical information and pedigrees. He had a few local informants and a sense of the local culture, though he did not have a professional collaborator who was an anthropologist or scholar of social medicine. A more important missing element in this proposal was an obvious, trackable genetic disease. There was no particular evidence that the Tangier islanders suffered from higher rates of genetic disease as a consequence of their geographical and social isolation. With the Amish, it was disease that first attracted McKusick to the group and disease that enhanced the productivity of his field research.

Organizing the Amish

As McKusick recalls these events, as he relates them in his own publications and to interviewers, in the fall of 1962 he happened to read a profile of a country doctor in Lancaster, Pennsylvania, David E. Krusen.[12] Krusen "indicated to the author of the article—in a paper in a pharmaceutical 'throwaway'—that achondroplasia is frequent among the Amish."[13] McKusick had been working with patients with Marfan syndrome and had an interest in disorders of connective tissue. Most dwarfing conditions involve a defect in connective tissue, and so the doctor's story interested him. But the rates of achondroplasia that the doctor described seemed too high. "Since true

MOMENTS OF TRUTH IN GENETIC MEDICINE

achondroplasia is a dominant with markedly reduced reproductive fitness, one would not expect it to have a high frequency in an inbred reproductive group."[14] McKusick suspected that this was some other condition.

A few months later, Penn State sociology professor John Hostetler submitted a book proposal to the Johns Hopkins University Press for possible publication. The book was to focus on the medical, social, and cultural beliefs of the Old Order Amish. One of those reading the proposal for the press was McKusick, who was then chief of the Division of Medical Genetics in the Department of Medicine at Hopkins. McKusick was intrigued and wrote to Hostetler. He heard that Hostetler would be visiting Baltimore "sometime this month." Could he give a talk on the Amish?

"We have had occasion to become much interested in blood groups and other physical anthropological characteristics of the Old Order Amish in Mifflin County," wrote McKusick, an interest stemming "out of an observation of an unusual type of hereditary disorder which seems to occur with relatively high frequency in this Amish group." McKusick had not yet been out in the field, but he explained specifically that he was interested in making "some arrangement to get blood samples on a representative group of individuals." He realized that acquiring these blood samples would be facilitated by a knowledge of "family structure, the attitude of the group toward illness and conventional medicine, etc., etc." He offered Hostetler a $50 honorarium and, as was often his custom with invited speakers, an invitation to stay at McKusick's home.[15]

The work that had attracted McKusick's attention was a sociological study of Amish culture, including Amish attitudes toward illness and Western medicine. For his analysis of medical ideas, Hostetler used two sources. First, he examined the detailed medical reports published in twenty-six issues of the Lancaster County newspaper, *The Budget*, in 1960. Second, he interviewed and surveyed forty-six physicians who worked with Amish patients in Ontario, Pennsylvania, Indiana, and Iowa. He found that the Amish, when they were in need of health care, favored both folk and Western medicine, seeing no contradiction between them. "The Amish find nothing in the Bible which would prevent them from using hospitals, dentists, fluoridation, surgeons or anesthetists" and "nothing sacred" about the "Amish healing arts." The frequent, detailed newspaper reports of illness, he found, were intended to provoke the Amish community to send letters to those who were ill and to visit particularly the chronically ill and aged relatives. In other words, the medical reports were intended to induce community action (Hostetler 1963, 1963–64).

These newspaper reports would make it easier to track down individuals suffering from particular diseases, including genetic diseases. For McKusick's purposes, the interviews with physicians serving these communities were even more important, for of these forty-six physicians, thirty-one "believed that the incidence of possible hereditary pathologies were greater among Amish than among non-Amish" (Hostetler 1963, 1963–64).

When McKusick eventually compiled a list of the qualities that made the Amish good research subjects, he included "great interest in illness," "clannishness," and a high consanguinity rate. The social practices in Amish populations around Lancaster and Mifflin counties in Pennsylvania produced genetic disease, he suggested, bringing it out into the open both biologically and culturally. The acceptability of cousin-marriage combined with a closed breeding population to make recessive genetic diseases more common in the offspring of these marriages. Meticulous Amish genealogical records facilitated the research. And the Amish practice of chronicling illnesses of all kinds in newspapers made disease socially visible and easier to track. The cultural specifics of this population seemed to be almost tailored to field research in human genetics, a fact that McKusick appreciated fully.[16]

In addition, the Amish, despite their isolation from mainstream life in Pennsylvania, were subject to the standard systems of vital statistics applied to all state residents. Their births and deaths were recorded; their death certificates, with the name of the attending physician and a cause of death, could contain relevant details for a pedigree. And the Amish, with a European background, used readily recognizable kinship categories that they shared with the investigators—not always a feature of population studies. Hopis in Arizona, for example, used *brother* to refer to at least four different kinds of blood or cultural relationship (C. M. Woolf and Dukepoo 1959). A "brother" in Lancaster County was a male sibling with the same biological parents as the speaker.

Within a few months McKusick was in the field, accompanied by two local guides: Hostetler, whose scholarly work on Amish culture he found so useful, and Krusen, the country doctor who had been the focus of the feature story. Krusen took McKusick to see the families in which he believed achondroplasia was present. Hostetler served as a hybrid culture broker, an

MOMENTS OF TRUTH IN GENETIC MEDICINE

academic and scholar who was born into an Amish family and maintained social ties to the Amish community. Characteristically McKusick pulled together various forms of local knowledge and created allies that could help him build an information-gathering network. Finding one's way through a social and cultural system—and through the country roads of Lancaster County—required cultural and medical informants.[17]

McKusick and his research team, which eventually included medical students and others, began the first phase of their research by asking physicians in these areas general questions about familial disease and the frequency of spontaneous abortions (McKusick, Hostetler, and Egeland 1964). This survey suggested that dwarfism was unusually common in the Amish, and by 1964, after a series of field trips, McKusick identified a particular form of dwarfism, Ellis–van Creveld syndrome, in thirty sibships and more than fifty individuals in the Lancaster County Old Order Amish. Mental retardation had been reported as a possible feature of this syndrome, but on the basis of these fifty cases, McKusick and his colleagues concluded that it was not.[18] With the data from one group, the total number of reported cases of EVC in the scientific literature doubled (McKusick, Eldridge, et al. 1964a).[19] The group also quickly found an entirely new genetic disease, a second type of dwarfism that was, like EVC, recessively inherited. Cartilage-hair hypoplasia, as they called it, was found in fifty Amish sibships, and it, too, was not accompanied by mental retardation (McKusick, Eldridge, et al. 1964b).

Over the next year, McKusick, Hostetler, and Janice Egeland (a recent Yale University Ph.D. who had written a dissertation in medical sociology on the Amish) conducted a more formal survey of five hundred physicians who worked with Amish patients, practicing in Amish areas of Pennsylvania, Ohio, Indiana, and Ontario. They asked physicians whether they had encountered certain hereditary anomalies. They also asked about specific reports of dead children or genetic disorders. "We would enclose a checklist of conditions such as albinism, dwarfism, congenital deafness, muscular dystrophy, mental retardation, cystic fibrosis and so on. In many instances the questionnaire was followed up by a visit to a doctor's office."[20]

Their first publication on the Amish as subjects of genetic study was published in the *Bulletin of the Johns Hopkins Hospital* in 1964. "Genetic Studies of the Amish: Background and Potentialities" was devoted almost entirely to the history and culture of the Amish. Group religious beliefs, educational

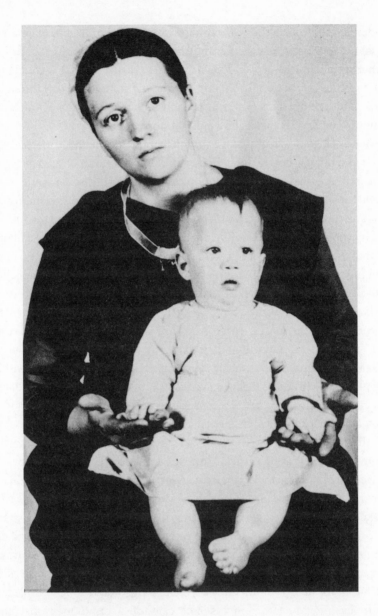

"The Amish Madonna," holding child with Ellis–van Creveld syndrome.
Courtesy of Dr. Robert Weilbaecher.

practices, and attitudes toward technologies were explored, but much of the paper dealt with the populational and genetic histories of the Amish. The authors explored the "range of overlap" between Pennsylvania, Ohio, and Indiana populations of Amish by comparing the last names of clergymen in these groups, and they found that this supported the idea that the Pennsylvania population differed from the Ohio and Indiana groups, which shared more last names (McKusick, Hostetler, and Egeland 1964, in McKusick 1978, 3–25).[21]

The Amish were also producing large families. Seven to nine children was the rule, and Amish parents did not stop having children after the birth of a child with a genetic disease. Furthermore, children with mental retardation or other disability were kept at home, which meant that they could be readily studied "in relation to the rest of the family." Populations with similar characteristics had been studied in Switzerland, Sweden, and other areas of Europe, but McKusick, Hostetler, and Egeland pointed out that the most informative such population in the United States was the Utah Mormons, who like the Amish were genealogically inclined, relatively immobile, clannish, and closed to outsiders (13).

The difficulties in "realizing the full potential of the Amish for genetic studies" were related to Amish suspicion of outsiders and reactions to some aspects of medical science. Amish families were in general reluctant to agree to autopsy: of the thirty-six deceased persons with EVC reported in the journal, only one was autopsied (14). The authors printed in their paper a response from a thirty-year-old Amish man with EVC who wrote to McKusick to say that he would not agree to be examined by the Hopkins physicians. "I feel I am exactly the way the Good Lord intended for me to be, even before I was born. So I feel no human hands or brains can do a thing about me or anyone like me, if it is the way the Lord wants it, no matter how highly educated anyone is. I am happy, have work, friends and can support myself. So what more do such people want?" (15). Another prospective participant in the study sent McKusick a postcard declining to participate. "I am not interested in going in the hospital so don't come around for me because I am not going in. And you don't have to stop in to see me either. I am allright there isn't anything wrong with me and I don't think much of those x-rays you want. So don't stop in to see me. I am not interested in your stopping by."[22]

In both cases respondents were contesting their status as objects of medical and scientific interest and as diseased. The first was satisfied that his condition was God's will; the second that there was nothing wrong with him. They would not participate in the medical research or in the construction of EVC—of their short stature and extra fingers—as a genetic disease. They were fine, and in any case God wanted things this way. And they were resistant to the medical technologies and interactions that McKusick's work would require: "I don't think much of those x-rays."

By their resistance and disengagement, these two men demonstrate the powerful and little-explored role of active community support in genetics and perhaps all biomedical research. They are the negative examples, the counterexamples, that testify to the dispersed nature of technical knowledge in biomedicine. What they knew of genetic disease could not play a role in any scientific paper because they were not talking. What their bodies "knew" in blood and tissue was similarly disbarred. It could not appear in any chart amalgamating data from multiple cases. By their refusal to participate, they bring into relief the acts of participation and cooperation of their peers: it is just as much a choice to choose to speak, to give blood, or to remember. Neither choice is of self-evident legitimacy, and both choices have consequences for scientific and medical research undertaken in any population.

Most Amish asked to participate were willing to do so, some of them even serving as informants and field-workers for McKusick's project, and many of them providing him with clues, leads, and suggestions for tracking other cases. They wrote to McKusick for advice about whether to marry distant cousins. They spontaneously reported abnormal births in their extended families. The "mother in sibship 23" reported two other cases that became sibships 25 and 26; the "father in sibship 3" was the "informant" for a case in sibship 10. McKusick was recording exactly who told him what, as participants informed the research group about their families, about dead grandparents and dead infants, about mental retardation and intelligence and extra fingers. The Amish were actively engaged in the construction of Amish genetic disease and apparently interested in the research and what it would reveal (McKusick 2000). This engagement emerged clearly in McKusick's data-collection network.

Assembling the Network

Over several years, McKusick built a sieve that could lift a specific, visible form of genetic disease out of a social network. He was trying to find all cases of short stature and extra fingers. He used public records, Amish genealogy books, birth reports, newspapers, health professionals, and Amish contacts. He began to subscribe to *The Budget*, which published reports about tonsillectomies, birthday parties, bad backs, and why someone missed church (a knee injury). He kept in touch with the undertaker who handled most Amish funerals. He surveyed the records of hospitals. He wrote to school officials to ask them to excuse students to attend examinations in his clinic.[23] He wrote personal letters to teachers, nurses, parents, and physicians. He was looking for polydactyly, extra fingers, evidence of which could be found in the state capital, Harrisburg, a decade after a neonatal death in Lancaster and could be remembered by a midwife attending a stillbirth, or by a grandmother, or a sibling. Evidence of disease that ended a life of only four hours could be seen despite the poor resolution of the record and the temporal and cultural distance between a midwife on an Amish farm and the head of the Division of Clinical Genetics at the Johns Hopkins University ten years later. Every baby mattered, including those who lived only twenty minutes. "Did this baby have extra fingers?" McKusick asked a nurse present at a birth in September 1969.[24] The entire population needed to pass through the sieve.

Parents provided McKusick with descriptions of their children's bodies, recalling the morphologies of stillbirths and neonatal deaths years after the fact so that those pathological forms could become a part of the pedigree. "Mother states no extra fingers. Four children living and well," he recorded in one case of a four-week-old infant who died; "Mother states extra fingers were present but apparently not of type in EVC," in the case of a thirty-three-year-old daughter who died in 1959, before McKusick began his study.[25] "Mother states that there were no extra fingers. First discovered heart condition at about 1 month," he recorded in the case of an infant who died at six weeks of age in 1959; "Mother states that there were 'no brains.' Anencephaly?" in a 1956 stillbirth.[26] Maternal descriptions could rule out EVC or certify its presence. Amish family members could also lead McKusick to other communities, as did one Pennsylvania father who said his siblings in Ohio had children with the same condition that affected his own. This father theorized that the condition was hereditary. "Now this may sound strange," he wrote to McKusick, "but five years ago my sister had a baby girl with the same trouble as ours. And a month or so later my brother's wife gave birth to a baby girl also with this same thing. Could it be hereditary?"[27] There was what might be called a folk epidemiology in the community itself, a network of knowledge and interpretation that could help McKusick identify relevant families and relevant bodily forms.

The Amish genealogy books encoded this folk epidemiology, and McKusick used them to construct pedigrees. In the fall of 1963, for example, he encountered an Amish family that had three adult siblings with polydactyly and chondrodystrophy, Ellis–van Creveld syndrome. The genealogy book of the family revealed that there were two other siblings who died young, one as an infant and one as a teenager. Did these children have the same traits? He made his inquiry to the family physician, but the family physician asked the father. The father reported that the infant probably did, for it was a "short, chunky little baby," but the teenager certainly did not and had died of pneumonia.[28] A genealogical text prepared for religious reasons helped make the family a scientific resource, and a physician consulted for his specialized knowledge simply asked the father, who diagnosed both infant and teenager. There were many kinds of knowledge in this reconstructed pedigree.

Local physicians were also an important resource, of course, and tracking down a new case often began with an appeal to the attending physician. In September 1963, for example, McKusick asked a Pennsylvania physician about a child who had died at Lancaster General Hospital in the spring of 1962 and who was reportedly born with signs of achondroplasia and heart problems. McKusick had probably learned about this death from the Amish newspaper, which reported the details of neonatal deaths, including the presence of abnormalities. He asked the attending physician about the size of the family, the presence of extra fingers, and the health of the parents. "I have been much interested in the last year in hereditary disorders among the Amish and have been making a particular study of dwarfism. I would appreciate any information you can give me on Amish dwarfs." He enclosed a list of cases and families he already knew of and asked the physician if he knew any others, closing with a proposal that he would drop by the physician's office on his next field trip. This particular physician cheerfully answered all the questions and invited McKusick to stop by his office.[29] Medical expertise joined with religious records, state bureaucratic forms, and the social experience of families as the pedigree took shape.

As McKusick's database grew, he recorded where and how he had learned of the existence and status of each person with EVC. The provenance of any given case included how it was ascertained initially, on what basis EVC was diagnosed, what the health status of the affected individual was, and how the pedigree was constructed in relationship to other pedigrees such as the Fisher genealogy. McKusick recognized fully the importance of this documentation, and he even included it as an appendix to his two-part 1964 paper "Dwarfism in the Amish." The "frequency of EvC as determined in this study is so unusually high and such a large proportion of the cases had died before the study was performed that it is deemed essential to outline briefly the features of each sibship, and to indicate the mode of ascertainment and basis for diagnosis in each case" (McKusick, Eldridge, et al. 1964a, 1964b, in McKusick 1978, 119). The appendix listed twenty-nine sibships, and a typical listing included a description of the affected child and some indication of the reliability of the information ("polydactyly is . . . absolutely certain in the minds of the Amish informants") (121). In sibship 11, the affected family member was reported to "work hard with horses on farm" (suggesting perhaps that he was relatively healthy), and in sibship 3 there was a chance meeting in a hospital corridor that led to ascertainment of the EVC diagnosis. For most of his cases, and certainly for his most important pedigrees, McKusick had many sources and all were recorded. A case might be documented in birth and death certificates in Harrisburg, a family Bible, hospital records, phone calls from a physician, letters from family members, personal observation on a particular field trip, or reports from a descendant, or recorded by one of his field-workers. As McKusick followed these signs he collapsed distinctions between sources of information, accepting as equivalent reports from midwives about recent births, from mothers about infants born a decade earlier, from physicians and nurses, and from a man reporting on the health and stature of his long-dead great grandfather. The pedigree pulled together many fabrics into a pattern that rationalized the heterogeneity of the sources.

On the Road

Road trips from Baltimore to Lancaster County, about a two-hour drive, were an important part of this field project. They seem to have been relatively casual. A small group from Hopkins would plan to visit families, some of whom would prove to be away from home. These visits involved brief medical examinations and casual information gathering. McKusick learned about deaths and births, school performance, and new business plans. The trips often included a visit to popular tourist spots to buy jam or eat lunch. They were tours of Amish heredity.

Victor McKusick and his twin brother, Vincent, made such a trip together in February 1975. Vincent was not a geneticist but a prominent Maine jurist. His presence on the trip reinforces the idea that it was the sort of thing one might invite a visiting relative to join. With a driver and a physician who worked in Lancaster County, the two brothers went to see Amish families. The four men made nineteen stops, apparently in a single day, although this is difficult to imagine. Their report of this field trip is an eclectic mixture of social and medical observation.

One man with symphalangism seemed "a very effective person" with a "comfortable home and pleasant family." At another home, McKusick reported, "I was bitten by a small dog." They learned about a recent death— "the funeral was being held the day we were there"—and frightened a young child who "sobbed throughout our time there." At another home they learned that the family was "no longer Old Order Amish." They visited a family in which three of the children had what they identified as "pseudomongolism." "They have an appearance that suggests mongolism. They are very affectionate . . . They are also short of stature. There is no simian crease in them." The family had "not been written up" nor had "a similar family of pseudomongolism" in the same community. They described a teenager with EVC, a "very fine boy" who was a junior in high school and had a job in a hardware store. "All the children work hard on the farm taking care of 27 milk cows." Another boy had been run over by a "wagonful of stones" but was now doing all right. "He sings a great deal and takes voice lessons. He is doing reasonably well in school."[30]

McKusick made a similar field trip in 1982, on a Saturday when there were no Grand Rounds. He and his team "seized the opportunity to visit Lancaster County," where they made sixteen stops. They "stopped by Kaufman's Orchard to get some apples and other things" and while there ran into one of the families they were planning to visit. After a few more visits to families, they came to the home of a "very attractive young couple" who were "prepared for having church at their house the following day. They expected 70 or 80 people. She had made 32 pies the day before!" They were the parents of a infant daughter with EVC who had died eight weeks earlier, when she was six weeks old. "This is a new EvC sibship," the McKusick group reported. Apparently based on this interview, they recorded some details of the baby's medical problems. The child had "polydactyly and a heart problem" and had not been considered a good candidate for heart surgery. The group

MOMENTS OF TRUTH IN GENETIC MEDICINE

next visited an Amish school, where they learned from a teacher that there had recently been a "discipline problem" at the school. "This is hard for us to understand. She said that what they call a disciplinary problem is different from what we would call a disciplinary problem. A disciplinary problem to them is refusal on the part of the boys to wear their hats when they are out on the playground, or a refusal to button the top button of their shirts!" At another home they found a blind father, a daughter with serious medical problems, and a "competent" mother—"She has to be to run a farm and raise a family with a blind husband. The husband makes brooms of several varieties and I bought a barn broom and a house broom from them." The McKusick group "ended up having a family style meal at the Harvest Drive Restaurant."[31]

These trip reports combined cultural observation, standard tourism, medical intervention, and genetic theorizing. The visitors from Baltimore were bemused by the school discipline problems, admiring of the woman with the blind husband, and sympathetic to the young couple mourning the loss of their daughter. They were purchasing brooms, eating family-style Amish food, and learning about children who had been run over by wagons.

The warm reception apparently accorded to McKusick and his team during these trips is remarkable given that McKusick was studying stigmatizing conditions, dwarfism and mental retardation, in a stigmatized population that was already explicitly in tension with mainstream American culture. Furthermore, these disease conditions were a consequence, biologically speaking, of the social choices that isolated the Amish and other plain people from their neighbors: recessive genetic diseases appeared in the Old Order Amish precisely because these Amish did not mix, socially or sexually, with the populations around them. When McKusick expressed his concern about this state of affairs, he framed it as a problem of research access rather than of further stigmatization: "It is important to observe that any publicity about mental retardation in the Amish would be tremendously detrimental to the further progress of this study, which is still far from complete."[32]

Indeed, McKusick did not particularly want the Amish who provided him with information and insights and with descriptions of their dead infants to see (or read?) the scientific papers he was publishing. In a letter in March 1965 to a Strasbourg, Pennsylvania, physician who was managing one of the EVC cases, McKusick outlined some of the problems that this patient was having and suggested that a cardiac catheterization would soon be necessary. He closed this clinical letter with the following: "I am enclosing a reprint of our Ellis–van Creveld paper. I, of course, do not want it to get into the hands of our Amish neighbors."[33] The article presumably referred to— the 1964 paper "Dwarfism in the Amish" published in two parts in the *Bulletin of the Johns Hopkins Hospital* (McKusick, Eldridge, et al. 1964a, 1964b)—included photographs of recognizable Amish people, both adults and children, who had EVC. Their names were not included in the text, but their faces were not blocked out and they would presumably have been known to many Lancaster County Amish. The combination of practices here, including the publication of recognizable photos and the expectation that such publications could be kept from the population under study, was unremarkable at the time.

I want to suggest that the Amish in all their material, intellectual, and emotional embodiment were both present and absent in the texts that McKusick and his confederates produced. The specific Amish subjects were visible, depicted in photographs, included in pedigree charts. They moved through the texts, interjecting insights, observations, and data points as the analysis unfolded, as the authors invoked Amish perspectives and described their roles. Yet they were not audience. The data that drew on their knowledge did not speak to them, nor was it intended to speak to them. They might benefit from the medical findings at the clinical level, the level of the body, but they were not expected to benefit as readers or thinkers.

There was another case pictured in one of McKusick's early papers, in the form of a reproduction of a seventeenth-century drawing of an elegantly posed human skeleton. The skeleton was that of a newborn infant, drawn standing, skull tipped quizzically, arms slightly raised, in a formal posture that would have been impossible for the living child. "Thrown into the river at birth," the newborn had been retrieved for its scientific interest. If not the focus of maternal love, it could at least be the focus of the love and desire that informed natural philosophy. It had seven digits on each hand, eight and nine digits on its feet, and some had theorized that it was a case of EVC. The Amish, with all their historical specificity, could be visually linked to a seventeenth-century Dutch newborn. Their disease bound them to a distant place and time.[34] The Dutch newborn was perhaps a record of the provenance of Ellis–van Creveld syndrome, a record of the "derivation and passage of an item through its various owners." The gene moved through human populations, leaving traces, signs, clues in the standard systems of

MOMENTS OF TRUTH IN GENETIC MEDICINE

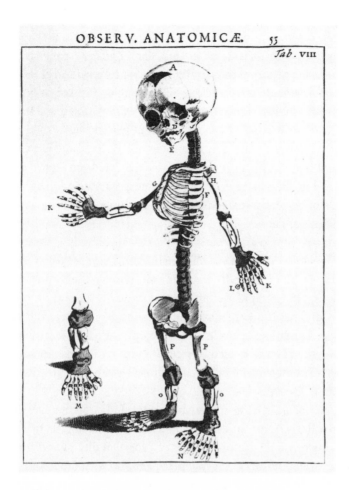

Skeleton of an infant, described in 1670 and reprinted in
McKusick's early papers on EVC.

accounting for people and recording their medical status. It was revealed
in the body of an infant pulled from a river and written in the birth records
of the state of Pennsylvania. The pedigree highlighted these clues, brought
them together, and situated them in a narrative that could make them scientific resources.

Cataloguing Disease

The well-known story of Ellis–van Creveld syndrome provides a rich case
study of population genetics and human genetic disease. But McKusick's

real claim to historical attention lies in his creation of a catalogue of human heredity, *Mendelian Inheritance in Man*,[35] and I want to explore now the development and history of this very different but related form of textual record that situates genetic disease. First published in 1966 but originating as a 1960 list of X-linked traits, *Mendelian Inheritance in Man* is a standard reference work widely used by many different parties with a stake in genetic disease or gene mapping. In its first iteration, it listed the names, descriptions, and citations for all genetic diseases that had been reported in the scientific and medical literatures. In its contemporary iteration, it is an online database of chromosomal locations and multiple links to related sites for each disease or gene.

It was and is also an expression of a way of thinking about genetic disease that is relevant to the rise of the Human Genome Project. It is a mapping project, and it provides mapped testimony to the imprecision of the phenotype: "Phenotypic overlap is not necessarily ground for considering [two conditions] fundamentally the same or even closely related" (MIM 1986, xx). Conditions with similar symptoms, affecting similar bodily systems, might in fact be caused by very different mutations or genetic abnormalities. Chromosomal location, like clinical expression, provided a way of organizing disease.

The acts of classifying and relating natural phenomena have long been understood by science studies scholars and anthropologists to express culture and to reveal the defining values of a social or disciplinary group. What belongs together in any given system depends on what properties are ranked higher and given priority. Genetic disease is sufficiently complicated that it could conceivably be categorized by the bodily system affected (skin, cartilage, cardiovascular), degree of disability (mild to fatal), level of pain experienced, potential for intervention, physical appearance, or biochemical and developmental pathways. It could also be categorized by frequency, rates by ethnic group, time of expression in the life course, age at which it can be diagnosed, method of testing or identifying, and so on. Not all of these classification systems would be equally utilitarian in terms of clinical management and scientific understanding, but they do all represent reasonable or imaginable ways of putting together and separating conditions understood to be genetic diseases. The question of the utility of chromosomal location could also be raised. What does knowing the location of a disease gene— its precise site on the genome—make possible? What does it accomplish?

Why is this way of understanding genetic disease seen as powerful or com-pelling? And why has the map, the location, become the primary means of distinguishing genetic diseases, when two diseases can be located very close together on a chromosome and yet have nothing else in common? Location on the genome does not necessarily have any bearing on the bodily experi-ence of disease: biochemical disorders are scattered across the chromo-somes. What, then, does location reveal?

McKusick's catalogues defined genetic diseases as points on chromo-somes long before such points could be specified. Even in 1960 they were locations (on the X chromosome, for example) and products of specific hereditary changes or errors. "Disease A and Disease B are either the same disease, if they are based on the same mutation, or different diseases." There was no room for the notion of a "spectrum of disease" in medical genetics, McKusick proposed, because diseases were caused by specific genetic changes and they could and would be strictly delineated (eventually) on this basis. This is what *Mendelian Inheritance in Man* could facilitate: the proper delineation of genetic disease. Mapping locations on the genome would eventually have an intrinsic value and provide a way of overcoming the con-fusion produced by examination of the phenotype. The phenotype was an untrustworthy guide to genetic events. It could mislead, suggest connections where there were none, or suggest genetic causes when in fact the causes were environmental.

McKusick sometimes denied that he was engaged in "butterfly collect-ing," expressing his sense that collecting was a less-than-respectable activity. McKusick certainly was, however, collecting with a remarkable passion and with an explicitly medical agenda that has been fully realized in the inter-national effort to map and sequence human genes, the Human Genome Project. He was an early and eloquent proponent of what I call the cata-loguing imperative: the increasingly powerful idea among medical geneti-cists that the compilation of a list (or, later, map) of genetic traits, birth de-fects, and diseases in human populations could transform medical practice and patient care. Cataloguing and mapping disease genes seemed to be the edge of a (desirable) slippery slope, the discovered gene leading gravita-tionally to therapy, the map and clinical intervention apparently linked, falling into place together. McKusick proposed the possibility of gene ther-apy in 1966 and the idea that all human genes should and could be mapped in 1969.[36] At the same time that he was working with Amish populations, making field trips, tracking down new sibships, he was imagining genetic medicine grounded in a complete guide to the text of heredity.

McKusick's first catalogue, compiled in 1960, was a list of traits found on the X chromosome. Traits on the X chromosome are expressed in a male even if they are recessive, because males have only one X chromosome. Most genetic diseases are believed to be recessive, and therefore traits found on the X chromosome are more visible than those found on other chromosomes (McKusick 1962). As McKusick and his colleagues began their work with Amish communities in 1962, they began to wonder exactly how many and what kinds of recessive conditions might be more common in genetically isolated populations. They wanted to know what they needed to be looking for. "What rare recessive disorders might one expect to encounter in inbred groups such as these?" (MIM 1966, vii). They compiled a catalogue of recessive conditions. By 1963, "the complexity of the recessive catalog prompted exploration of computer methods for assembling, revising and indexing" (vii).In 1964, McKusick's group turned to the creation of a catalogue of dominant traits then known in the scientific literature. Supported by two National Institutes of Health grants, in 1966 the first edition of *Mendelian Inheritance in Man* was published.

McKusick laid out a justification for the catalogues of both dominant and recessive traits in his first introduction, in 1966. The catalogues needed to be explained and justified because they required "very considerable effort" to assemble—indeed, they consumed much of the rest of his professional life. By the 1980s, producing *Mendelian Inheritance in Man* was a cottage industry, with a product in constant revision in response to scientific reports and new interpretations. What justified this labor? First, he said in 1966, genetic counseling demanded accurate diagnosis and familiarity with the "experience reported in the literature." Phenotypic similarity could lead to mistaken diagnoses and poor counseling. Those engaged in telling prospective parents what their risks were for bearing a child with any particular disease must be able to recognize the disease properly so that their advice conformed to what was known in the literature more generally. The catalogue, then, was a guide to genetic counseling.

But it had other purposes as well, as McKusick suggested in his opening essay. Genetic disorders "give us insight into the normal." The catalogues were "like photographic negatives from which a positive picture of man's genetic constitution can be made." And all knowledge of "man's ge-

MOMENTS OF TRUTH IN GENETIC MEDICINE

netic constitution" could be assumed to be useful "in the long run." The "numerology" of the catalogues suggested that much of the genome was unknown and many more genetic diseases were waiting to be discovered. When genetic disorders were examined closely, he noted, they often turned out to be more than one entity. "What at first is thought to be one entity is found to be several clinically similar (i.e. phenotypically similar) but fundamentally (i.e. genotypically) distinct disorders" (MIM 1966, ix). The genotype was fundamental, and the clinical experience of the disease was a misleading guide to its meaning. McKusick made the comparison to bacteriology: just as a wide range of disease conditions had been vaguely characterized before the bacteriological revolution and then found to be caused by specific microorganisms, so, too, many genetic conditions were vaguely characterized and poorly understood in the absence of an explicit genomic location and mutational cause. Finding the gene or its location pinned down the disease and gave it a substantive reality that the vagaries of the clinic could not supply.

The entries in this first version of the catalogue were organized alphabetically by the name of the trait. "Red Hair," entry 2449, merited four citations and a short paragraph describing key observations about the inheritance of red hair (MIM 1966, 230). The entry for Schilder's disease questioned whether it was a proper category at all: all cases reported were "probably in fact" three other diseases (234). Schizophrenia required a discussion both of the genetic data and the nature of the condition, which might be manic-depression and might not be a single entity at all (234). A quick examination of the entries as a whole suggests that part of the work was related to nomenclature. Diseases were called one thing yet sounded like another, more than one name had been applied to the same condition, and different conditions had been called the same thing. McKusick was clearing out the underbrush of naming and classifying, providing a single source for names and diagnoses that could rationalize the published literature on genetic disease.

By 1966 the catalogues were on magnetic tape, a form in which computer information was then stored, so that they could be easily revised and republished as new information accumulated. McKusick said he planned to maintain and republish the catalogues indefinitely. "I have no illusions of either the infallibility or the completeness of these catalogs," he noted. Errors and omissions were likely, and honest differences of opinion might lead readers to see classifications as inadequate or flawed (MIM 1966, x). But this was an ongoing project, with room for revision.

In the seventh edition, in 1986, McKusick summarized the growth of his own enterprise. The number of loci identified in 1966 had been 1,487; twenty years later it was 3,907. The number of citations of certain journals and the number of journals consulted had also grown. Pediatric journals were heavily represented as sources for the 1986 edition, but also important were journals in neurology, hematology, and ophthalmology (MIM 1986, xiii). The 1986 edition also included a much expanded and much more complex discussion of the phenotype, the genotype, and the problem of identifying genetic disease. McKusick pointed out all the ways that conditions could mimic genetic diseases, including "Munchausen syndrome by proxy," a form of child abuse that can affect two or more children in a family and therefore seem genetic. He mentioned kuru and slow viruses, congenital infection, and abnormalities in the children of mothers with phenylketonuria. And he discussed molecular genetics and the sequencing of mitochondrial DNA by Fred Sanger and his collaborators at Cambridge University (xxii). He included an appendix on the human gene map, based on eight international workshops on human gene mapping between 1973 and 1986 (xxxix). By 1986, human molecular genetics was a practical reality, the chromosomes were being mapped, the database was available online, sequencing was a technical possibility, and the vagaries of the phenotype could be resolved by genotypic analysis—analysis of DNA itself.

The Amish work was implicated in McKusick's decision to create a catalogue of "rare recessive phenotypes in man." This catalogue was described in a 1964 paper in *Cold Spring Harbor Symposia on Quantitative Biology* by McKusick, Hostetler, Egeland (all involved in the Amish field research) and Roswell Eldridge, who at the time was a fellow in medical genetics at Johns Hopkins and later worked on neurological problems with a hereditary basis. The catalogue included two hundred rare phenotypes. Its development as a single source for genetic diseases was justified by the fact that reports of a form of dwarfism associated with very fine, white hair in the Amish led other investigators to notice similar cases in Finland, Italy, and France. "Previously the cases were too few to impress persons with the fact that hair and cartilage changes are associated in a hereditary syndrome" (McKusick, Hostetler, et al. 1964, 62) A catalogue, then, could help geneticists recognize rare genetic diseases when they saw them. The practical problems of understanding disease in Amish populations led McKusick to catalogue known genetic diseases, and the growth of his catalogue proceeded in tandem with the growth of medical genetics as a discipline.

Conclusion

Finding a particular gene has required extracting biological material from someone, convincing individuals to contribute some portion of their body and some portion of their personal history to science. The blood and the narrative could then be embedded in a larger narrative, a pedigree documenting family history, a causal model clarifying the nature of the genetic defect based on inheritance patterns revealed in the pedigree, an origin story about the source of the mutation based on its population distribution, or, later, a map upon which a particular disease could be placed in relation to all other genetic diseases. The potential inscriptions of the field work were many, complex, cumulative. But the basic inscription, the first point of translation, was the pedigree, and producing a pedigree was unquestionably social work.

Recent work in the history and sociology of science has explored the properties of the field sciences with special attention to the "chronic issues of status and credibility that derive from the social and methodological tension between laboratory and field standards of evidence and reasoning." Field work entails phenomena that are "multivariate, historically produced, often fleeting and dauntingly complex and uncontrollable," as Henrika Kuklick and Robert Kohler (1996) have noted. Given its complexity and uncontrollability, the field seems to be almost unsuited to the production of scientific knowledge. Kuklick and Kohler observe that "it may seem astonishing that any robust knowledge comes out of field work. Yet it does, abundantly and regularly" (3). They also point out that in the field, unlike in the private space of the laboratory, scientific work is shaped by the social interactions of professionals, amateurs, and local residents whose cooperation is both necessary and rarely acknowledged. These social interactions were and are a critical part of constructing the pedigree. Genetic and familial histories must be assessed in collaboration, with the aid of memory, state records, family stories, and biological tests. Pedigrees are rich amalgamations of many ways of knowing. And despite its difficulties, the pedigree was and is the "cornerstone" of medical genetics.[37]

In a scribbled, undated set of notes in McKusick's papers, he outlined his concerns about pedigree methods. First, he said, genetic diseases involved a "dynamic process." "At the time you study them you can't give the last word." A disease might unfold as a person aged, and human beings, McKusick pointed out, live as long as geneticists do and were therefore outside the normal time frame for scientific study. Second, he said, there was a "lack of specificity" in clinical manifestations. The trait was "far from the gene." Reading down to the hereditary material from the clinical sign was difficult, complicated, sometimes not possible. McKusick fully appreciated the complexity of moving from the body to the genotype. Finally, there was the question of familiar seeing. He knew that the more you looked at anyone, the more you saw. The clinician in a standard examination might miss something relevant to understanding the disease and therefore the pedigree. "The more individual[s] personally studied and studied from a specific point of view, the greater the reliability."[38]

In a 1977 grant proposal, McKusick said his project would permit him to document "new recessive disorders among the Old Order Amish." He could also gain insights into "incompletely characterized Mendelian entities" by studying this population. There were at least sixty cases of cartilage-hair hypoplasia, for example, and some new reported cases of chondrodysplasia with severe combined immune deficiency. There were also a "presumably autosomal recessive form of osteogenesis imperfecta," many cases of cleft hand and foot, and six sibships with Kaufman syndrome.[39] The Amish were a rich treasure trove of genetic disease, and McKusick's field trips provided access to these diseases, which were called to his attention in casual conversations at the market and in trips to the local school. The fragments of detail—a child born dead, a sister with mental retardation— became part of grant proposals. The "medical tourist" brought back enough information to secure funding for another trip.[40]

A grant reviewer, considering a proposal to the National Institutes of Health, wrote that "the type of research is more descriptive than innovative, but an enormous amount of data which makes a significant contribution to clinical genetics results from astute clinical observation and intelligent interpretation of the results. It is in this field that Dr. McKusick is a world leader."[41] As this assessment suggests, McKusick's own relationship to both clinical medicine and high science mirrored the complicated interplay of the clinic and the laboratory in the rise of medical genetics. McKusick occupied a critical borderland between science and medicine. When he was nomi-

MOMENTS OF TRUTH IN GENETIC MEDICINE

nated for membership in the National Academy of Sciences, one skeptic asked, "But is he a scientist?"[42] And was human genetics a science?

In the 1970s, McKusick kept a dwarfed miniature poodle, Vanilla, who was one of two dwarfed puppies in a litter from the breeder Rebecca Tansil. In March 1973 Vanilla was bred back to her father, Can.Ch.Andeches Ready to Go. The pregnancy was expected to be risky, with a potential need for cesarian delivery. McKusick made provisions for Vanilla to be closely watched at the Division of Laboratory Animal Medicine at the Johns Hopkins University School of Medicine, under the care of Edward C. Melby, professor and director of this division. The poodle would be taken to the laboratory, not the farm, two or three weeks before expected whelping. There, she could be watched more closely for problems.[43] McKusick's family pet—his companion animal—was an animal model for the condition he was studying in human populations. Vanilla's pregnancy was handled in one of the world's finest medical schools, and she was used to produce, through inbreeding, a biological state that mimicked Ellis–van Creveld syndrome. McKusick was an overseer of dwarfing conditions, a manager of bodies and genomes and catalogues, and a perceptive student of the pedigree, both canine and human, in all its biosocial complexity.

Like other human geneticists of his generation, McKusick was struggling to find a way to study human populations and human genetic traits that mitigated the chaotic conditions of the (human) field. Genetic field work was the study of phenomena that were deeply disordered and uncontrollable. Managing this disorder required a keen attention to the histories of people, blood, families, cultures as they intersected in a single phenotype. It required following a malformation of the limbs and an abundance of fingers through the many texts, stories, and memories where they were documented and translating these signs into an authentic record of heredity that could presume to hold its own in the laboratory. Creating the impeccable pedigree was one of the great achievements of postwar human genetics. Making its contingency functionally invisible—or irrelevant to its usefulness as a guide to the genome—was a social project of enormous complexity.

It is not a coincidence, I think, that many of the most prominent practitioners of field research in human genetics have also been practitioners of their own distinctive form of historical reconstruction. James Neel wrote an autobiography (1994) and many self-reflective works about the discipline and about human history. The population geneticist William Schull published a memoir (1990) of his work in Japan. McKusick has published several historical reconstructions of the rise of medical genetics and the history of the Human Genome Project. He has also published scholarly articles in historical journals on other medical topics (McKusick 1949, 1953, 1992, 1996). He has shown an intense historical sensibility in other ways as well, for example in his preservation of virtually all of his own papers and in his work with the archives at Johns Hopkins. Arno Motulsky, who studied inherited hemoglobin anomalies in central Africa, and Bentley Glass, who studied a religious sect to explore questions of genetic drift, have both published on the history of their field, Glass has worked with archival institutions to preserve records relating to the history of genetics (Motulsky 1971; Glass 1988)

These powerful and influential scientists have constructed narratives that encompass not only their own fields but human history in general. They are the monitors of the human genome, tracking its biological qualities and its scientific interpretation through time. And the two endeavors—writing history and writing medical genetics—may be intertwined. In Vassiliki Betty Smocovitis's study (1996) of the evolutionary synthesis, she explores how the architects of the synthesis sought to build a compelling grand narrative that could bring together under a single canopy all scientific studies of living things. For many biologists, the chaotic biological sciences seemed to lack a unifying explanatory frame. The synthetic theory was therefore both a way of explaining the mechanisms and forces involved in the evolution of life and a reflection of the dreams, desires, and practices of twentieth-century biologists.

Like the architects of the synthesis (though on a smaller scale), the architects of postwar medical genetics needed to bring together a diverse range of practices, technologies, and materials from many different fields and to situate these in a single, coherent narrative. I propose here that the field studies of specialized populations were important places for the construction of this narrative. Geneticists sought to explain why particular diseases persisted in human populations, and they repeatedly tried to find diseases that, like sickle cell anemia, were preserved because of a heterozygote advantage in a given environment.[44] They also used field studies to explore the biological history of the human species as revealed by the quirks in the genome that inbreeding or isolation made visible.

It is conventional to attribute the Human Genome Project to the high technologies that made it practical, but I here attribute the project to the embrace of the cataloguing imperative by physicians and scientists who encountered genetic disease in the field, in their research, and in clinical practice. Collecting genetic diseases around the world, wherever they might spring up, was an elaborate exercise in pattern making with a broad medical agenda. The population work of the 1950s and 1960s tracked the role of heredity in susceptibility to infectious disease in human populations, in cancer, obesity, and many other complex human traits. And the idea that more knowledge of human genes was critical to an understanding of all human disease and all human qualities was a central justification, both political and scientific, for the mapping project. To locate the origins of the Human Genome Project solely in technological feasibility, as so many commentators have done, is to overlook the powerful ideology of the mapping project as it was carried out in remote regions of Africa and among socially isolated groups in Pennsylvania and as it was reflected in McKusick's massive catalogue, *Mendelian Inheritance in Man.*[45]

CHAPTER 4

[SQUASHED SPIDERS]

Standardizing the Human Chromosome and Other Unruly Things

In 1961 the University of Glasgow pathologist Bernard Lennox noted that the British medical journal *Lancet* had recently been "freely littered" with images "said to look like masses of squashed spiders" (Lennox 1961). The spiders in question were highly processed human chromosomes, shaped roughly like Xs. Either photographed or drawn for journal publication, they appeared in papers detailing the medical effects of chromosomal abnormality. In a new laboratory process, they had been retrieved from human blood: the red cells removed with phytohemagglutinin (an extract of the red bean *Phaseolus vulgaris*), and the white cells incubated in plasma and synthetic tissue culture medium, frozen in midmitosis, and, with "judicious squashing," flattened out in a single plane to be photographed.[1] This method produced images that were a triumph of human cytogenetics (Ford and Hamerton 1956).

Lennox's essay in the *Lancet*, "Chromosomes for Beginners," was a primer for the practicing physician. Lennox proposed that the human chromosome had lately become an object of medical interest: through recent innovations in human cytogenetics, classical genetics was finally making its way into medical practice. "At least we can now really *see* something," Lennox noted. Although the chromosomes were "only a corner of the field," that corner was now illuminated by "a refreshing beam of good, honest morbid anatomy, something tangible, unequivocal, simple to understand." The human chromosomes were the known cause of "four well-established and reasonably common syndromes and a great number of rarer variants, less well-studied." And thanks to the deliberations at the Denver Conference on Nomenclature the previous April (1960), the chromosomes could now be distinguished from each other, even though they did seem to look very much alike. Each human chromosome now had a number, and the scientific community had agreed to recognize and abide by this numbering system. The average reader of the *Lancet* (presumably a physician), Lennox proposed, would find that it was "well worth making the small effort to catch up with the main lines of discovery" in this field. Human cytogenetics, he proposed, had become something medical practitioners needed to understand (Lennox 1961).

From a perspective less respectful to the practicing physician, the German-born medical geneticist Klaus Patau, then at the University of Wisconsin, portrayed medical interest in human cytogenetics as a distraction. The "recent explosive development of human cytogenetics" had attracted the "enthusiastic interest" of many physicians who were heretofore ignorant of cytogenetics, he said in a 1963 review. And while this interest was laudable, it had also led to an "outpour of publications . . . that had more than the ordinary admixture of dilettantism." Anyone trying to keep track of this literature, he said, had to cope not only with rising volume but also with being "perpetually challenged to judge the validity of the observations and the good sense, or otherwise, of the interpretations." Many review articles produced to explain this broader literature to "outsiders" were "not written by specialists and added to the confusion rather than helped." And "even the specialist faced a formidable task when he tried to make a comprehensible story of the chaotic and rapidly expanding field."[2] From Patau's perspective, human cytogenetics exploded on the medical scene in the late 1950s and quickly attracted a storm of interest from persons whose disciplinary training was inadequate to the task at hand.

These accounts suggest that human cytogenetics was embraced by medical professionals in the 1960s in a chaotic and confusing period of technical and intellectual change. A transformation in laboratory practices in the late 1950s made human chromosomes easier to see and assess, and elucidation of the connections between chromosomes and health made them objects of medical attention.

The subjects of this interest were twenty-four X-shaped objects, visible in the nucleus of the cell only during certain stages of cell division: the twenty-two pairs of autosomes and the two sex chromosomes, X and Y. Accurately counted only in 1956 and not conclusively distinguished from each other until the 1970s, in the 1960s the chromosomes were both medically important and easy to misidentify. They were therefore the focus of intense debate.

In this chapter, I explore the emergence of human chromosomes as quantitative signs of disease. The "unruly things" of my title included—besides human chromosomes that looked alike—karyotypes that were unclear, patients who resisted the painful sternal puncture, and physicians and scientists who did not quite follow all the rules of standardization established at the Denver, London, or Chicago conferences. My story explores the social and technological process of scientific standardization, specifically the standardization of human chromosomes before banding technologies greatly simplified the problem in the 1970s. But I am interested more generally in the role of standardization in science and medicine. I use *standardization* here to refer not only to the rules for naming and depicting chromosomes but also to the rules for assessing the intuitive knowledge of the experienced cytogeneticist, submitting tissue samples to data banks or providing genetic counseling. I refer, then, to the negotiated agreements that undergird genomic medicine. Recent historical works have explored the role of standardization in the cultural authority of science. Standards of measurement were often linked by their developers to timeless, trans-historical truths and to divine order, though to the modern historian they look quite specific to a particular time and place (Hunt 1994; Wise 1995; Schaffer 1994, 1995). Some historians have interpreted standardization as a critical part of professionalization, knowledge production, and economic value. Precise measurement matters to the public, to commerce, and to the state. Metrology was useful science, and it engaged the attention of some of the most prominent figures in the history of science. Standardized measurement is a critical resource, not just a statement about nature. It facilitates exchange of knowledge, goods, technologies, and ideas (Hessenbruch 1999).

In a series of papers, for example, the historian of science Simon Schaffer has proposed that standardization practices and debates provide insights

MOMENTS OF TRUTH IN GENETIC MEDICINE

into the negotiated authority of scientific knowledge. He explores Lord Rayleigh's development of electrical resistance standards in the early 1880s, showing that the values of electrical resistance, "which were designed to be independent of anyone's individual reputation," were, ironically, bound up with Rayleigh's own high reputation for accuracy and personal reliability (Schaffer 1994). From a different perspective, the historian Ted Porter proposes that numerical standards are valued for their impersonality at least as much as for their apparent relationship to natural truth. Numbers, he suggests, leave behind individual experience, and idiosyncratic experience was the focus of increasing suspicion in the late nineteenth century (T. Porter 1996). There is also a small but growing literature on standardizing biological things, including living things such as mice (Rader 2004), rats (Clause 1993), and flies (Kohler 1994).

My case study here intersects with this earlier work on several levels. My actors are not, of course, dealing with a phenomenon, such as time, that can be divided in many different ways. The chromosomes are specific objects, quite capable of asserting their own individuality as they appear in the processed cell (I am here adopting psychological language to describe the actions of a cellular body, but this is strictly a literary device). The objects to be named are biologically significant and clearly differentiated, and while the exact length of railroad ties might be arbitrary in some senses, the identification of specific chromosomes was (ideally) not arbitrary.

The transformation of genetic disease in the postwar period was facilitated by the standardization of diseases, chromosomes, cell lines, techniques, and other entities that often defy easy categorization or that change through time. Biological things are messy, variable, and changeable. The "general problem of the heterogeneity of biologically active substances" (Cockburn et al. 1992, 2) played out around tuberculin antigen in the 1930s, blood products in the 1940s and later, diphtheria and tetanus toxoids in the 1950s, and polio vaccines in the 1950s and 1960s. The World Health Organization (reluctantly) took up the job of preparing international "requirements for biological substances" in 1949, and the organization's role in vaccine standardization continued to expand as vaccination played a larger role in its own programs. At the same time, the International Association of Biological Standards was founded in 1955 in order to bring together state interests, corporate scientists, and research workers interested in the control and standardization of reliable biological products.

But scientific and medical experts effectively devise "good enough" categories that can facilitate intellectual exchange. The labor involved in achieving this is sustained and enduring. Standardizing genomic objects and practices continues into the present, the subject of annual meetings and many publications. If one of my goals is to understand the labor through which genetic disease came into medical prominence, this labor of standardization must be taken into account. The international pressure to develop standards for naming and depicting chromosomes after 1959 was a consequence of their increasing medical importance. But in plain words, many chromosomes looked very much alike, even to trained observers. Chromosomes were therefore critical medical resources that were "resistant" to the standardizing gaze.

The rules that came to govern the karyotype were neither arbitrary social constructions nor reflections of absolute biological truth. They were shaped by convenience, professional priorities, technological capabilities, individual actors, and normal errors. They were historical products in every meaningful sense of the word. The rules facilitated the equation of clinical disease with abnormal chromosomes and, by extension, with other, still invisible, chromosomal qualities. It made genetic disease medically accessible, which in turn brought it to the attention of the Centers for Disease Control (which sought to regulate cytogenetics laboratories in the 1970s) and many state legislatures (which became interested in the public health implications of chromosomal disease). After prenatal diagnosis became a part of health care in the United States in the 1970s and 1980s, chromosomes became objects of interest to medical insurers, for obvious reasons (Rapp 2000).

The technical experts in my story have taken the process of naming and distinguishing chromosomes as seriously as I have. In his engaging first-person history of cytogenetics, the Chinese-born University of Texas cytogeneticist T. C. Hsu described the "sense of relief, as if a historic document was being written" when the report of the Denver Conference on Nomenclature was completed. Hsu went so far as to compare this report to the U.S. Constitution (Hsu 1979, 61). The British cytogeneticist John L. Hamerton reprinted in full the standardization reports of the Denver, Chicago, and London conferences in his 1971 survey of the field, and he even included additional commentaries and amendments on standardization and chromosome measurement (Hamerton 1971, appendices). Certainly between 1956

MOMENTS OF TRUTH IN GENETIC MEDICINE

and 1970, standardization engaged the sustained energy of many of the leading figures in medical genetics and cytogenetics and was extensively discussed in major journals, including the *Lancet, American Journal of Human Genetics, Cytogenetics,* and *Nature,* not only in formal papers but in letters and editorial comments. Standardization justified three international meetings in six years, and it was coupled by participants to the making of reliable scientific truth.

The human chromosomes could eventually be reliably characterized anywhere in the world. Those working in Kuala Lampur or other exotic sites could purchase scientific kits that would make the chromosomes visible ("just add blood").[3] And the peculiarities of chromosomes could be abbreviated in specific, arcane formulas—46,XX,t(Bp–;Dq+) or T(2s–;18l+), for example.[4] A code that could present chromosomal events in a compact and readily understood formula was put in place, at great effort, over a period of about a decade. In 1956 the human chromosomes were barely differentiated from each other. By 1970 they were distinct, stable objects of sustained technical and medical attention and the locus of a wide range of disease states. Chromosomes became visual markers of pathology and critical images in a new, broader conception of genetic disease.

Seeing the Human Chromosomes

Investigation of the number of human chromosomes began in the late nineteenth century. The chromosomes were so difficult to see that early counts varied widely. The German biologist D. Hansemann found normal human tissues with 18, 24, and more than 40 chromosomes (cited in Hamerton 1971, 2). Other cytologists examining human cells reported from 16 to 36 chromosomes, though there was some consensus that human beings had a total of 24 chromosomes. The Belgian cytologist Hans von Winiwarter counted between 46 and 49 chromosomes in his samples in 1912 and concluded that males had 47 chromosomes and females 48 (Hamerton 1971, 4).

Drawing on recent work in the chromosomal determination of sex (especially the work of Nettie Marie Stevens), Winiwarter proposed that human females had two X chromosomes and human males had one X (and no Y). Several other researchers followed up on the proposal that human males had no Y chromosome, sometimes finding a Y and sometimes not. University of Texas cytologist T. S. Painter in the early 1920s counted from 45 to 48 chromosomes in testicular biopsies from three incarcerated mentally

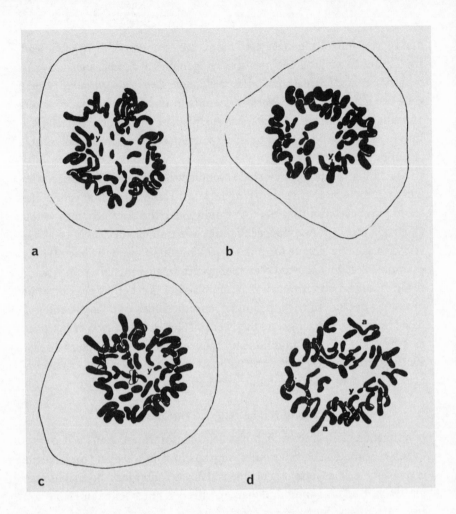

T. S. Painter's renditions of human chromosomes, 1923.

ill patients, noting in a 1921 paper that the clearest plate he had so far studied had 46 chromosomes. Painter found a Y chromosome; his work was respected, and the idea that human beings did not have a Y chromosome was abandoned. In 1923 Painter came to the conclusion that 48 was the correct diploid number in humans, and for about the next thirty years other cytologists saw the same number, or at least tended to believe they had miscounted if their numbers did not match Painter's (Martin 2004; Kottler 1974; Hsu 1979; Painter 1921, 1923).

Tjio and Levan's ideograms. From *Hereditas* 42: 5 (1956). Published with permission of Blackwell Publishing.

For Hsu, looking back at his own reluctance to challenge Painter's numbers, this period of relative stability and consensus, from 1923 until 1956, was frustrating evidence of the complacency of practitioners in the field of cytogenetics. "It is somewhat irritating to find that the cytologists of that era were apparently satisfied with their work, since so few of them sought technical improvements" (Hsu 1979, 8). Hsu himself recalled "forcing" forty-eight chromosomes from his prepared slides. Yet it is also easy to sympathize with Vogel and Motulsky, who observed in 1986 that the "jumbled masses of chromosomes in old illustrations make us understand the difficulties encountered by the pioneers who tried to count human chromosomes" (20). Cell preparation before the wide use of the "hypotonic shock technique" produced chromosomes stacked atop each other in an almost indistinguishable mass. The assumption that the number was known intersected with the confusing products of prevailing tissue culture techniques to validate Painter's erroneous 1923 count. By 1955 the fact that humans had forty-eight chromosomes had been confirmed many times by many different investigators (Kottler 1974). But in 1956, that number changed.

Modern human cytogenetics is conventionally interpreted as beginning with two 1956 papers, the first by Indonesian cytologist Joe-Hin Tjio and Swedish plant cytologist Albert Levan and the second, confirming the first, by cytogeneticists C. E. Ford and John L. Hamerton, both at the radiobiological research unit of the British Atomic Energy Research establishment at Harwell, near Oxford (Kevles 1985, 238–40). These two papers reported, for the first time, conclusive counts for the chromosomes in the human cell. The critical number was forty-six: twenty-two pairs of autosomes and the two sex chromosomes, X and Y.

Tjio, of the Aula Dei Experimental Station at Zaragoza, Spain, and Levan, of the Institute of Genetics at Lund, Sweden, were working together in the summer of 1955, looking at human tissue culture methods. Levan had been experimenting with Hsu's 1952 use of hypotonic solution to plump up and spread out the chromosomes to make them easier to see.

Hypotonic solution is a mixture of salt and distilled water that causes the chromosomes to separate from the spindles during mitosis: the cells swell and thus the chromosomes can disperse. It was first used in cell culture by two women scientists in the 1930s, one working with chick embryos and the other with grasshopper embryos (Lewis 1932; Slifer 1934). This method was almost ignored by cytologists working with human cells, until it was inde-

MOMENTS OF TRUTH IN GENETIC MEDICINE

pendently rediscovered by Hsu. While working with human cells at the University of Texas medical branch in Galveston, he had tried adding distilled water to rinse the tissue specimens before fixation (Hsu 1952; Kevles 1985, 240–41).

Levan, who drew on Hsu's methods, was interested in cancer. He had come to the subject through a circuitous interest in chromosomes in onion root tips, which were often damaged, and in similar damage seen in chromosomes in cancer cells. By 1955, his appointment in Sweden was in the Cancer Chromosome Laboratory, and he had for several years been working with mammalian cells, examining both tumor cells and normal cells. Early in the summer of 1955 he found that pretreating cells with both colchicine and hypotonic solution produced better images of the chromosomes. In August, Levan's long-time collaborator in plant cytology, Tjio, arrived in his lab.

They planned to apply the new techniques that Levan had worked out to human cells and quickly acquired four very small human embryos from recent legal abortions, courtesy of one of their colleagues in the Virus Laboratory of the Institute of Bacteriology at Lund. Suitable human material was thus readily available in the context of a major research institute in Sweden, and embryos, numbers 1 through 4, provided 261 cells in which the number of chromosomes could be counted. The familiar structure of twentieth-century biomedicine may make this ready access to suitable human tissue seem unremarkable, but it is not. Acquiring the right kinds of human biological material for work in human cytogenetics was not so simple in 1956 (see below). A blood sample or cells from a cheek swab would not have provided the kinds of images they needed, for technical reasons that I explore later. Tjio and Levan were fortunate in their institutional setting and in their relationship to the Virus Laboratory where the fetuses were acquired.

They thought Hsu had overdone the hypotonic solution in 1952. "We consequently tried to abbreviate the hypotonic treatment to a minimum, hoping to induce the scattering of the chromosomes without unfavourable effects on the chromosome surface" (Tjio and Levan 1956a, 723). They also modified the squashing. For chromosome counts "the squashing was made very mild" in order to keep the metaphase chromosomes in one cell together and intact. In many cases single cells were squashed under the microscope by slight pressure of a needle.

Their vivid explanation of how they squashed cells, described in their published paper (Tjio and Levan 1956b), provides an evocative image of physical labor and tacit knowledge. "Mild" and "thorough" squashing were categories apparently understood by their readers, who would be other cytogeneticists presumably aware of the distinctions. Squash techniques were decades old in plant cytology. But plant cells are tough and must be squashed vigorously, whereas mammalian cells had to be managed more carefully and gently to avoid breaking the cell. Hsu proposed that some people, those with thumbs that do not bend upward, for example, simply could not manage a proper squash (Hsu 1979, 36).

Tjio and Levan reported that they were "surprised to find that the chromosome number 46 predominated in the tissue cultures from all four embryos, only single cases [cells] deviating from this number" (Tjio and Levan 1956b, 1). Lower numbers were frequent, they said, but always in cells that "seemed damaged." There could have been some squashing errors as well, when "one or two solitary chromosomes [were] pressed into a 46 chromosome plate at the squashing." But apparently there were only 46 human chromosomes, Tjio and Levan concluded, even if an occasional cell with more or fewer showed up. "With previously used technique it has been extremely difficult to make counts in human material," they suggested. But cytogeneticists' expectations also seemed to have led them to see 48 instead of 46, even when the techniques were sufficient for an accurate count. "For example we think that the excellent photomicrograph of Hsu published in Darlington's book . . . is more in agreement with the chromosome number 46 than 48 and the same is true of many of the photomicrographs of human chromosomes previously published" (5). Despite this bold claim, the next paragraph expressed modesty. "We do not wish to generalize our findings into a statement that the chromosome number of man is $2n = 46$, but it is hard to avoid the conclusion that this would be the most natural explanation of our observations" (6). In Britain, Ford and Hamerton read this paper and promptly backed it up with their own observations (in their case on meiotic preparations from testes) (Ford and Hamerton 1956). Other confirmations followed. By 1958, cells from seventy-three other individuals had been found to have 46 chromosomes, like the four fetuses at Lund. On the basis of the analysis of cells derived from these seventy-seven bodies, the number of human chromosomes became 46 (Tjio and Puck 1958).

Understanding the importance of this change requires some familiarity with a few technical details. Chromosomes can be subjected to cytogenetic analysis only when they are condensed enough to be seen. This condensation takes place only in the metaphase stage of mitosis in dividing somatic cells or in meiosis, a type of cell division occurring only in germ line or gonadal cells. But whereas gonadal cells or specialized somatic cells such as bone marrow divide frequently and can be caught in the act more easily, the more readily available cells in blood or skin are slower to divide and more difficult to "catch" in the process. At the time, the materials that did contain appropriate, rapidly dividing cells required painful procedures to extract and therefore required high levels of patient cooperation. Cells in the bone marrow, spleen, or lymph system, for example, are good subjects for cytogenetic analysis and do not require artificial stimulation to produce cell division in vitro. But getting such materials out of a patient involves biopsies or even surgery. A large systematic study of human chromosomes that required a painful clinical interaction was simply impractical. Patients would not tolerate it. The techniques that made it possible to use human blood were therefore important for both technical and social reasons: physicians and scientists needed patients' cooperation to acquire bodily materials, and blood was much easier to get than bone marrow.

Patients' resistance and the resulting tissue availability can shape an entire field. Hsu noted that after Jerome Lejeune's characterization of Down syndrome as being a consequence of an extra chromosome, "funny looking kids" and people in mental institutions became prime targets of inquiry. "Funny looking kids" was a loose diagnostic category used by geneticists in private correspondence, and geneticists encountering such children began to karyotype their chromosomes to search for abnormalities.[5] But Hsu noted that inmates of mental institutions and criminal institutions also "contributed heavily to our knowledge of human cytogenetics" (Hsu 1979, 41). Hsu's use of the active verb, *contributed,* is particularly apt, for human subjects are routinely, if perhaps invisibly, active producers of natural knowledge of disease. Clarke Fraser, who was then on a National Institutes of Health Genetics Study Section, remembered a period in the early 1960s when the National Institutes of Health was flooded with applications for funds to study chromosomes taken from asylum inmates, people at institutions for the retarded, prisoners, and malformed children. "Cytogenetics laboratories were bombarded with requests for chromosome studies on achondroplasia, hemophilia, schizophrenia, what have you."[6] As these applications suggest, human cytogenetics changed when chromosomes could be acquired without the painful extraction of bone marrow or testicular materials. Large populations could be bled and studied with relative ease. The sheer volume of karyotypes produced new knowledge, and the technological interventions that made a normal blood sample a suitable material for chromosomal study made possible the extremely rapid development of knowledge of human chromosomes in the 1960s.

In this early period, almost every human cytogenetics paper I have encountered cited a paper in *Experimental Cell Research* that described a method for the preparation of white blood cells cultured from "human peripheral blood," a standard blood sample, the "the most easily obtained tissue" (Moorhead et al. 1960). Standardization of the human chromosomes thus extended to the types of materials taken from human bodies and the procedures to which these materials were subjected. (See Hamerton 1971, 8–30, for a detailed discussion of cytogenetic techniques.) It was precisely this ability to standardize—chromosomes, techniques, protocols—that led to the recognition and understanding of the nonstandard chromosomal variation in human populations. As cytogeneticists looked at more human chromosomes from more populations, they found unexpected connections between odd chromosomes and complex human disease.

Chromosomal Disease

The idea that some human syndromes might be caused by extra chromosomes or by chromosomal aberrations had been proposed as early as 1932 by the Dutch ophthalmologist P. J. Waardenburg, a specialist in the study of inherited eye disease. It had also been proposed by an American pediatrician, Adrien Bleyer (Hsu 1979, 38–39). Waardenberg, who knew of chromosomal aberrations in plants, asked, "Why should it not occur occasionally in humans, and why would it not be possible that—unless it is lethal—it would cause a radical anomaly of constitution? Somebody should examine in mongolism whether possibly a chromosomal deficiency, or non-disjunction— or the opposite—chromosomal duplication—is involved. My hypothesis has at least the advantage of being testable. It would also explain the possible influence of maternal age" (quoted in Vogel and Motulsky 1986, 20). This sort of hypothesizing about causes and disease is commonplace in science and medicine and perhaps is only retrieved retrospectively when it proves to be correct. The comment does at least suggest that chromosomal disease might have been noticed decades earlier if technologies for human chromosome preparations had been more reliable.

By the late 1950s several investigators were interested in the possibility that Down syndrome was the consequence of a chromosomal abnormality. In 1959 the French cytogeneticist Jerome Lejeune, with Marthe Gautier and Raymond Turpin, counted chromosomes in fifty-seven "perfect" cells taken from nine "mongoloid" children. In these cells, they found forty-seven chromosomes instead of forty-six. The extra chromosome was small, and Lejeune and his colleagues considered it to be the cause of Down syndrome, then called mongolism. They announced their discovery in several publications, in both French and English (Lejeune, Turpin, and Gautier, 1959; Lejeune, Gautier, and Turpin 1959a, 1959b, 1959c; Lejeune and Turpin 1961).

Lejeune's group suspected they were seeing an extra, normal chromosome, but they thought it might also be a fragment of another chromosome. They postulated an explanation for the link to maternal age. "It is known that in *Drosophila* non-disjunction is greatly influenced by maternal aging," and the extra chromosome, if that was what it was, might be the result of nondisjunction of "a part of small telocentric chromosomes at the time of meiosis." In their extremely short English-language paper (less than two full pages), with a minimum of fuss over theory and technique, Lejeune and his group laid out a compelling explanation for a complicated and common human syndrome (Lejeune, Gautier, and Turpin 1959b). They proposed that many manifestations of Down syndrome, which affected the body as a whole, were caused by a small extra chromosome.

Over the next twenty-two months, at least twenty-three new human chromosomal aberrations were reported by various laboratories, a "rate of about a syndrome a month" (Lejeune and Turpin 1961). These included links between chromosomal aberrations and some poorly defined clinical entities that the chromosomal condition basically defined: the chromosomal difference permitted some conditions to be extracted from the undifferentiated mass of things causing mental retardation. "Individualization has resulted from the chromosomal findings" (175). Scientists found trisomies of chromosome 17, of "one of the big acrocentric chromosomes (group 13–15)," and of "a member of the group 16–18." One group reported an abnormal chromosome 13 in two cases of Marfan syndrome (Tjio, Puck, and Robinson 1959). Three new sex constitutions were described, including the human "super female" (XXX) (Jacobs et al. 1959), a "hypofemale" with one normal X chromosome and one partly deleted X (Jacobs et al. 1960), and a "curious constitution" of 48–XXXY in two men with Klinefelter syndrome

(Ferguson-Smith, Johnston, and Weinberg 1960). According to Lejeune and Turpin, "the desire of human geneticists to map the chromosomes of our species has been greatly enhanced by these new results." Unfortunately a cytological map was "still remote, mainly because of the impossibility of detecting stable and recognizable structures in the chromosomes themselves." At least, they noted, it was now possible to "use the standard system of nomenclature agreed upon in Denver last spring" (Lejeune and Turpin 1961). As Lejeune and Turpin articulated here, the nomenclature and the diseases were part of the same system, and an understanding of the diseases would depend in part on the adoption of a standard nomenclature.

By 1959 several different ways of naming and classifying the chromosomes had appeared. Tjio and Levan, in their 1956 paper, divided the chromosomes into three groups, M, S, and T (1956b, 4). Lejeune himself simply began naming the chromosomes he was seeing, identifying them as G1, Md1, Vh, Vs, C1, and so on. Other cytogeneticists looking at human chromosomes were numbering them; for example, Tjio and Puck in 1958 assigned numbers to all the chromosomes, though not apparently by size (18 came immediately after 12). Ford, Patricia Jacobs, and L. G. Lajtha in Britain divided the chromosomes into three groups by size and centromere, the centromere being the constriction or crossing point of the X shape of each chromosome. Some chromosomes are about equally long on both sides of the centromere, and some are unequal. The centromeric index is a measure of the ratio of the shorter arm to the overall length of the chromosome (Ford, Jacobs, and Lajtha 1958). Meanwhile Levan and Hsu grouped the chromosomes in "seven categories to promote easy identification," claiming that "broadly speaking, our grouping is the same as that used by earlier writers" (Levan and Hsu 1959).

In early 1960, Ernest Chu prepared an idiogram, an idealized human chromosomal complement, that he thought was in "almost complete agreement" with other idiograms published by other researchers: the "only major difference is in the systems employed in numbering the chromosomes."[7] The need for a unified system to avoid confusion and to serve as a working basis was apparent. "Hope was generally expressed, during recent discussions with a number of these workers, that a uniform nomenclature will soon be adopted" (Chu 1960a).[8]

As these varying interpretations and numbering schemes multiplied, the physicist turned biologist Theodore T. Puck at the University of Colorado at Denver called a summit meeting to resolve them. This was the April 1960 Denver conference. Participants in Denver focused not only on questions of nomenclature but also on visual depiction. The chromosomes as biological *and* pictorial objects were negotiated at this meeting.

The Denver Conference on Nomenclature

Puck arranged the conference at the suggestion of the British cancer researcher Charles E. Ford of the Radiobiological Research Unit at Harwell. The meeting was supported by the American Cancer Society, and participation was limited to "those human cytologists who had already published karyotypes."[9] These included cancer researcher David Hungerford of the Wistar Institute in Philadelphia,[10] cytogeneticist Patricia Jacobs of the Group for Research on the General Effects of Radiation at Western General Hospital in Edinburgh, Lejeune, Levan, Tjio, and Hsu (then at the M.D. Anderson Hospital and Tumor Institute in Houston).[11] Also invited, as neutral moderators, were three prominent geneticists who were not working in chromosome research, radiation geneticist H. J. Muller, *Drosophila* and human geneticist Curt Stern, and geneticist D. G. Catcheside. Puck apparently anticipated the possibility of unresolvable arguments. The moderators—three, so that a tie was impossible—could help resolve them should they occur.

The meeting was necessary because the "rapid growth of knowledge of human chromosomes [had] given rise to several systems by which the chromosomes are named. This has led to confusion in the literature and so to the need for resolving the differences." Following principles of "simplicity and freedom from ambiguity," the new system of nomenclature should be "capable of adjustment and expansion to meet the needs of new knowledge of human chromosomes." Over four days that included "many heated disagreements" (Hsu 1979, 59), the group sought to establish some concordance between the six systems then in use in the scientific literature. They decided to number the chromosomes 1 through 22, by descending size, even though in several cases the chromosomes were so similar in size that choosing which came next was difficult. The sex chromosomes, they decided, should be neither numbered nor named. They were to be just X and Y, a number for these chromosomes seeming to be a "superfluous appellation." The group also agreed that the twenty-two autosomes could be classified by size into seven groups, within each of which identification of spe-

cific chromosomes was not always possible. This was an important conces-
sion, and one that became in practice the primary "naming" method for several years.

There were other standards to consider. Measuring a chromosome involved taking a photograph of chromosomes after they had been frozen in
midmitosis, blowing up the photograph to 2,500× enlargement (defined by
microphotographic technologies), and using calipers or other measuring devices to assess overall length and the lengths of long and short arms. There
were proposals as early as 1966 that the measuring used in this process
could be automated and computerized, but the "skilled cytologist" remained
the repository of information that computers did not have (see below) (C. S.
Gilbert 1966).

In their report, published in both the *Lancet* and, as a letter to the editor,
in the *American Journal of Human Genetics* (Denver Study Group 1960a,
1960b), participants in the Denver conference supported the desirability of
"using a uniform system for presenting karyotypes and idiograms, but recognizing that individual variation in taste is involved, rigidity of design was
thought undesirable. However, it was recommended that the chromosomes
should be arranged in numerical order, with the sex chromosomes near to
but separated from the autosomes they resemble. It is desirable that similar ones be grouped together with their centromeres aligned." While the
choice between different schemes of nomenclature was "recognized [to be]
arbitrary," "uniformity for ease of reference is essential." With this in mind,
"individual preferences" had been "subordinated to the common good in
reaching this agreement." All members of the study group had agreed to use
the Denver nomenclature, and the group recommended that "any one who
would prefer to use any other scheme should, at the same time, also refer
to the standard system proposed here" (Denver Study Group 1960a, 1065).

Even those who disagreed with the Denver nomenclature, then, were
asked to follow it. The system would be useless without broad compliance.
Order could be imposed on the chromosomes only if order could be imposed on the behavior of cytogeneticists and other scientists. As things
turned out, one of the most prominent members of this select group had already misidentified an important chromosome, and his error was in effect
institutionalized, so that the last two human chromosomes have the "wrong"
numbers. Lejeune's group mistakenly identified the extra chromosome in
Down syndrome as the second smallest, chromosome 21, and Down syn-

MOMENTS OF TRUTH IN GENETIC MEDICINE

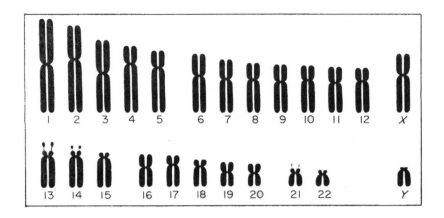

Denver conference numbering system. From *The Annals of Human Genetics* 24: 319 (1960). Published with permission of Blackwell Publishing.

drome became known as trisomy 21. As later imaging techniques made clear, the supernumerary chromosome in Down syndrome is in fact the smallest autosome and is therefore, according to the strict logic of the naming system, chromosome 22. But Down syndrome, perhaps the most well-known chromosomal disease, was already known as trisomy 21, and so this minor numbering anomaly persisted, by general agreement. This agreement was certified at the second Denver conference, in 1970, "when the extra chromosome in the Down Syndrome was officially designated as chromosome 21" (I. H. Porter 1974).

Patau was characteristically disturbed: "There was never any reason to call the mongolism chromosome 21, except for a very short period when it was believed that chromosomes 21 and 22 could be defined by the presence or absence of satellites, and when one or two cases of mongolism became known in which satellites had been seen on three G chromosomes. The identification of the mongolism chromosome as the satellite-bearing No. 21 very soon collapsed when it was found that satellites could occur on all members of this group. We usually follow priority in science, and priority has it that the smaller chromosome is No. 22. Mongolism should therefore be called trisomy 22" (Patau 1965, 72). Instead of revising the name of the disease, however, cytogeneticists renamed the chromosomes. An error was incorporated into the chromosome standardization process, and a "famous" chromosome retained its incorrect number because it was already so well known.

Predictably, almost as soon as the Denver conference ended, its conclusions were questioned. The intense interest in chromosomes produced new interpretations and new frustrations with both the chromosomes themselves (so difficult to manipulate and see) and with the system of naming them. Perhaps the most openly and consistently frustrated participant in this debate was Patau, who had not been invited to the Denver conference (he was included in the next meeting, in London) and who in a series of sometimes cantankerous papers and talks in the 1960s provided a lively running critique of the standardization process (Patau 1961).

The Denver report, Patau said, which had been "published and republished extensively," reflected not only a "sensible consensus on nomenclature as a matter of convenience but also an amiable rather than critical attitude of compromise toward contradictory data." If the term *standard* implied "nothing but wide acceptance" then the Denver system would qualify, he suggested. "However, the term implies more—at the very least clarity as to what is defined and what is not." Some of the conference statements, he said, seemed to contradict each other; for example, the appeal to further knowledge before more chromosomes were specifically named was in conflict with a later expectation that the numbers currently assigned would be "permanently fixed." The report reminded him of "blue and red filter[s] which, when combined, result in opacity" (Patau 1961, 933). He went so far as to suggest that the quantitative methodology of human cytogenetics was "as time honored as it is misleading. The fact that [two publications] support each other fairly well merely illustrates that repeated applications of a method with built-in systematic error will tend to mislead in the same direction" (Patau 1960, 250). The "whole thing seems to be a kind of 'newspeak'" (Patau 1961, 934).

Particularly dangerous, Patau said, was the claim by the conference that with "very favorable preparations distinctions can be made between most, if not all, chromosomes," for it had induced workers in the field "to publish karyotypes in which every chromosome is given a number—after all, who wants to confess that he has no 'very favorable preparations'?" Whereas some cytologists took the position that "much of the numbering of chromosomes that goes on has no particular meaning but . . . by and large no

MOMENTS OF TRUTH IN GENETIC MEDICINE

particular harm is being done either," Patau foresaw serious problems that were explicitly historical. "What is going to happen when an unambiguous definition of the one or the other chromosome becomes possible? Are future students to memorize the date at which a number that had been used haphazardly before acquired meaning?" (Patau 1961, 934). In a passage that legitimated Patau's concerns, a *Lancet* paper in 1961 by Albert de la Chapelle, of the Minerva Foundation Institute for Medical Research in Helsinki, blithely identified one chromosome as chromosome 7 and then went on to state "it could just as well have been called a 6 or an 8" (de la Chapelle 1961, 461).

Many other geneticists also questioned whether human chromosomes could or should be numbered at such an indefinite stage in their scientific history. As one paper put it, the "degree of accuracy in human chromosome identification leaves much to be desired and there is a great need for more exact description" (Ferguson-Smith et al. 1962). In practice, cytogeneticists continued to publish papers that clumped the chromosomes together by alphabetical designation—under the letters A through G, with A containing what would later be known as chromosomes 1 and 2 (the largest chromosomes) and C containing the troubling midsize group, chromosomes 6 through 12 and the X chromosomes. There seemed to be general agreement that chromosomes 4 and 5 were indistinguishable and, indeed, that many other chromosomes were indistinguishable as well.

In 1963 the burgeoning weight of these uncertainties led cytogeneticists to hold another international conference, this one sponsored by the Association for the Aid of Crippled Children in New York: the connection between standardized chromosomes and human disease was explicit in the funding source. The London Conference on the Normal Human Karyotype of 1963 was called to "consider developments since the Denver Conference and to assess the degree to which these may have aided characterization of individual chromosomes or may have revealed inadequacies in the earlier identifications." Those attending included the London anthropologist N. A. Barnicot, two representatives of the biology division of the Oak Ridge National Laboratory, and the more familiar cast of Lejeune, Ford, Jacobs, Patau (invited the second time, if not the first), Penrose, Puck, Stern, Tjio, and Hamerton (then at the pediatric research unit at Guy's Hospital Medical School, London). This group of distinguished scientists and physicians thus included individuals with a primary or institutional allegiance to genetics, cytogenetics, pediatrics, radiation biology, and even anthropology. The chromosomes had to be stabilized not only around the world but across these many disciplines.

The brief London conference report, which appeared in the new journal *Cytogenetics*, noted that cytogeneticists were still referring to the human chromosomes either by their numerical designation or by their assignment to groups, clustered together by size under the letters A through G. Size determined the group assignment of a chromosome, so the small Y chromosome was grouped with chromosomes 21 and 22 in group G. Conference members emphasized that "assignment of numbers to chromosome pairs on the basis of morphological criteria here discussed includes an element of arbitrariness; it does not ensure homology" (London Conference 1963).

In 1966, in Chicago, a third cytogenetics conference essentially established a standardizing ritual that continues into the present. As the structure and mechanics of the cell and the gene continue to be elucidated, there are always new problems of naming and classification to be resolved. Even by the time of the Chicago conference, the realms to be standardized had already begun to expand. Human cytogeneticists needed more than standardized chromosome names. They needed international standards for communicating information about tissue samples, rules for describing abnormalities, and even rules for carrying out field research. Proper procedures for cell culture and the management of population studies needed to be articulated and included in any registry report, and there needed to be general agreement about how populations could be studied.[12]

The nomenclature committee concluded that it was "important in a cytogenetic report that the procedures which were used to secure" materials be "clearly specified." Information should be included, the group said, on cytological techniques, scoring methods, and methods of ascertainment (13). The data deriving from human memory, the statements of those from whom blood had been taken, were part of what anyone assessing human chromosomes needed to know. The stories of families, of miscarriages and early heart attacks and malformed feet, were a necessary frame for the cytogenetic analyses. The way that a particular body came into the system (ascertainment) was important to the biological meaning of the chromosomes that it yielded.

The group sought consensus on the standards for recording this anamnestic data and the clinical findings. "Many reports that have appeared in the past might have been more valuable if they had included more com-

MOMENTS OF TRUTH IN GENETIC MEDICINE

plete data. Certain data are essential in all human cytogenetic studies and the interests and potential value of such studies increases [*sic*] in proportion to the completeness of the combined clinical, biochemical, genetic and cytogenetic evaluation of relevant individuals" (19). The phenotype, detected in the clinical examination, needed to be described as completely as possible, with direct reference to all known features in any previously described syndrome. If the association of certain developmental and anatomical traits were already known, then the presence or absence of each feature should be clearly stated. "In the case of previously unrecognized syndromes a full clinical and phenotypic description is required. This applies as well to the finding of a new or previously unrecognized chromosome abnormality" (20). The reports describing these clinical traits should avoid the use of "ambiguous terms," and eponyms should be used only "in conjunction with a clear definition." The recommended source for terminology was the International Classification of Diseases, World Health Organization (20).

Despite the relatively large number of people whose karyotypes had been completed by 1966, there were some individual patients who were "famous" and to whom references were commonly made in the scientific literature. A set of standard rules for referring to human subjects was therefore also explored, though not resolved, at the Chicago conference. Clear identification was important "at the present stage of human cytogenetics, so that where the same subject is ascertained or referred to in publications on more than one occasion, this can be recognized." At the same time, the privacy of patients needed to be protected. The intent was to devise conventions that permitted "the greatest possible information consistent with the protection of the family from public recognition" (22).

Finally, the uncertain future of genetic technologies made it advisable that other bodily materials be collected and stored. Serum, whole blood, and fibroblasts should all be collected and stored if possible. In terms of the needs of a planned chromosome registry, the method used for short-term blood culture must be precisely indicated. Those transmitting samples to the registry should make clear the type and volume of the inoculum, the culture volume, the type and concentration of mitogen, the formulation of media, incubation temperature, culture interval, type and interval of hypotonic treatment, type of fixation, and method of preparing slides. For cell cultures obtained from solid tissues, "whenever possible more than one primary culture should be established and biopsies should be made from more than one site." When preparations were made from living sources, the type of tissue and cytological methods should be specified.[13]

The abnormal chromosome could be properly interpreted only if situated in several frames: in a standard laboratory preparation method, in a pedigree, in an ethnic group, and in a clinical history. Every chromosome should be subject to the same techniques and accompanied by and embedded in the same information. This was standardizing on a grand scale.

The Standardized Field of Vision

Familiar seeing plays a critical role in many kinds of knowledge. It is particularly important in scientific fields, such as cytogenetics, that involve ambiguous visual messages. Judgment was a part of technical knowledge production in cytogenetics, and just as some cytogeneticists sought to standardize how cell cultures were handled and how the chromosomes were named, others sought to validate the judgment calls that permitted this naming and standardization. This work proposed that the intuitive powers of the human eye should be backed up by the impersonality and objectivity of numbers. It proposed an equivalence between different kinds of knowledge and different ways of seeing. What the cytogeneticist knew through intimate experience, the computer would know through more reliable means.

I want to explore here two of these studies of computers and individualized knowledge, because I think they present a wonderful tangle. The projects I consider are restless and uncomfortable with the fuzziness of familiar seeing. They express a critical tension over the unruliness of biological things and the uncertainty of human perception. Both the chromosomes and the mind observing them posed problems of trust. Neither could be counted on entirely. And the resolution of these problems was a technical one: the computer could validate both the reality of the cellular bodies and the reality or trustworthiness of human visual expertise. A machine's moment of truth could testify to the reliability of human perception.

In 1969, with support from the U.S. Atomic Energy Commission, a Boston physician and a Rochester, New York, computer specialist attempted to show that "there is an objective quantitative basis for classifying normal human chromosomes into eight distinct and identifiable groups" (Neurath and Enslein 1969). They proposed that such classifications depended on "an experienced cytogeneticist's visual pattern-recognition faculties, which are at least partly subjective." Experienced cytogeneticists might use features

MOMENTS OF TRUTH IN GENETIC MEDICINE

that were "hard to define and quantify" in sorting chromosomes, but those features were nonetheless actually present.

The researchers in this study began by measuring chromosome length, considering the short arm, long arm, total length, and centromeric index in various combinations. This "objective" quantitative measure they interpreted as a replication of the mental process of the experienced cytogeneticist. When karyotyping, they proposed, the skilled cytogeneticist "probably considers and solves a k-dimensional problem" (where k = an unknown, large number of parameters) with variables including arm shape, presence of satellites, presence of secondary constrictions, and so on. Yet, despite the subjectivity of the cytogeneticist, his or her mind was capable of multiple, simultaneous comparisons that could be verified and validated by mathematical analysis. "The advantage of a multi-dimensional over a two dimensional analysis in separating different classes of objects can be appreciated intuitively." But a mathematical computer program could produce "numerical results" that would "give objective backing for the cytogeneticist's conventional methods." The mathematical calculations were apparently constructed in this particular study as reassuring support for the cytogeneticist's more intuitive analysis, rather than as a challenge to conventional methods. The calculations would not overturn knowledge acquired by familiar seeing but would justify it. The numbers were not an alternative form of knowledge but a sign that the knowledge already accessible through the human retina was reliable and genuine (Neurath and Enslein 1969).

In the same year, a study by Dorothy Warburton and colleagues at Columbia University tested the ability of five observers to detect very small deletions in chromosomes, just by looking. The point of this study was to suggest that the deletions were real and of medical significance. The authors proposed that because the deletions could be detected by human observers under controlled conditions, the observed abnormalities were responsible for a syndrome "with features suggestive of a deletion of chromosome 4 or 5, but with grossly normal chromosomes"(Warburton et al. 1969, 97). This approach linked a particular configuration of the body (a clinical syndrome) to a subtle quantitative difference in the chromosomes.

The study began after six patients with characteristics linked to deletions of either chromosome 4 or 5 came to the attention of the Columbia group. The six patients had mental retardation, microcephaly (small head size), and a retarded growth rate. Yet, when their blood was processed and manipulated so that the chromosomes could be inspected, their chromosomes seemed to be normal. There were no obvious deletions or rearrangements to which the symptoms could be attributed.

Warburton and her colleagues decided that deletions were probably responsible for these abnormalities, because they had seen these same symptoms in patients with visible deletions. There was apparently a lengthy internal wrangle over whether deletions could be present or not. "We would like to emphasize the great difficulties we had [and] the previous disagreement among the observers" (106). In the end the group decided to test both their causal theory and their own visual skills. Could the deletions, if they were there, be detected by the human eye? "A blind study was carried out by five observers [the authors, identified by their initials in the published table] who were very familiar with human karyotypes, on a series of photographs from controls, patients with obvious deletions of the B-group and cases being tested for minute deletions." The photographs were identified only by number and were shuffled together randomly to be examined. Each observer "independently scored cells for the presence of a deletion of a B-group short arm."

In all, the Columbia group carried out three separate blind trials on 462 cells. Control photographs came from the "files of patients in whom there was no reason to suspect a B-group abnormality." They found that, in general, their scoring for the controls, major deletions, and minor deletions was consistent. They all tended to see deletions in some cases and no deletions in others. Consensus became evidence of reliability. They concluded, diplomatically, that each of them working independently would have come to the same conclusion. "There were no observers who were 'better' at discriminating between controls and cells from cases with a minute deletion, and a relatively inexperienced observer seemed to be as proficient as those who were more experienced in studying human chromosomes" (106). Seeing the chromosomes, they seemed to suggest, might not even depend on experience. Even those with limited experience might be able to detect slight deletions.

What was the ultimate purpose of this familiar seeing? It was prediction. The ability to recognize these small deletions was important, the authors suggested, because the deletions could reveal "whether the parents of a child with the clinical features of a B-group deletion are heterozygous carriers of a translocation and therefore likely to have more abnormal children" (106). The purpose of this calibration of the eye, the visual image, the statistically measured chromosome, and the malformed child was to detect the abnormal human body *before* it appeared. The cytogeneticist could see future children in a minor deletion of the short arm of chromosome 4, and his or her mind was an instrument that could move from the patient with abnormalities to the subtle deletion and then back again to the (future) child.

Standardizing the human chromosomes required quantifying their physical dimensions and quantifying the mental processes by which they were distinguished. It required establishing rules about naming and organizing on the printed page and rules about what kinds of stories had to accompany a piece of a human body that came into a cytogenetics lab. At great effort, a system was built around the human chromosomes that permitted them to serve as reliable visual markers of human disease.

Conclusion

For much of the history of human genetics, most human chromosomes could not be related to disease. Traits and diseases were understood to be hereditary in the late nineteenth century, with no reference to chromosomal location. The chromosomal determination of sex and the interpretation of hemophilia and other diseases as sex-linked constituted the only knowledge of human chromosomes as specific sites for disease or bodily traits. For much of the twentieth century, human geneticists had not even counted the number of human chromosomes properly. That most basic of scientific acts, quantification, faltered until 1956.

After 1956, human chromosomes were the focus of sustained, aggressive interest on many levels. For a brief period, human cytogenetics "developed into the most popular branch of human genetics." The research conducted on human chromosomes illuminated "the cause of many previously unexplained malformation syndromes." Visual images appealed strongly to physicians and to many biologists. The "surging popularity" of clinical cytogenetics was "all the more remarkable since during the first decade almost no practical significance of these results for medical therapy or prevention, apart from diagnosis and genetic counseling, seemed to be in sight." This "changed dramatically" when prenatal diagnosis became possible (Vogel and Motulsky 1986, 24).

Human cytogenetics thus played a critical and little-explored role in increasing medical interest in genetic disease during this period. New techniques of preparation made chromosomal disease *visible* both in the patient's syndrome and in the extracted cell. Genetic disease before this time could be seen or detected in the pedigree, the historical reconstruction of a family's history, and in the clinically abnormal body. But chromosomal disease had a specific cellular marker, an extra, missing, or abnormally shaped chromosome. The ability to literally see the physical cause of a complicated syndrome, to *count* it in that most elementary form of scientific observation, was particularly gratifying to both geneticists and physicians. When British geneticist Lionel Penrose saw the chromosomes of a patient who had both Down and Klinefelter syndromes, he found "the photograph of the cell from the man with two extra chromosomes, from which the intelligence level, the behavior, [and] the sexual characters can be confidently predicted, just about as astonishing as a photograph of the back of the moon" (quoted in Kevles 1985, 248).[14]

By 1974, karyotypes had been prepared based on samples from about five hundred thousand people, many of them institutionalized in prisons or mental hospitals (I. H. Porter 1974, x). Some authors barely acknowledged that the biological materials involved in these studies came from a human body or any sort of body at all. But many were quite specific, explaining that the patient was a "ten-year old male child" with a "mongoloid mother" from whom "a small piece of skin, less than 1 cm. was biopsied on May 23, 1959" (Levan and Hsu 1959). The included details about the people from whom blood was taken suggest what aspects of the human being seemed relevant to chromosomal events. Were the bodies British? Executed criminals? Had the person from whom the blood was taken been depressed? Did he or she have relatives who were diseased in some way? Various publications during this period included all these kinds of information about persons who were the subjects of cytogenetic studies. Most obviously, all such reports included sex and bodily source of the material (testicles, buccal smear, blood). But they also included details about race, nationality, psychological state, legal status, and parental status.

Of Painter's 1921 work, for example, Chu said that it was "interesting to note that the material used in his studies were [sic] testicular biopsies from three (two Negroes and one White) insane individuals" (Chu 1960b, 97). Painter himself had included the detail that the testicles were removed from these three inmates of the Texas State Insane Asylum in Austin because of "excessive self-abuse" (Painter 1923). Giannelli and Howlett reported in 1966 that they had acquired their material from a "phenotypically normal woman heterozygote for a D/D interchange." Another group of researchers stated that two of their subjects "had been hospitalized for neuropsychiatric reasons" and one had tuberculosis (Hungerford et al. 1969). These details about the body, mind, behavior, and illness suggest what kinds of problems cytogeneticists expected might be relevant to the chromosomal images they produced.

As karyotypes from thousands of human subjects were examined and compared, cytogeneticists found significant variations in the size and shape of chromosomes. They found that some cells had too many chromosomes and others did not have enough; that some chromosomes varied in size among individuals or even races (the large Y chromosome of Japanese men, for example, was the subject of several papers);[15] and that some had visible satellites or secondary constrictions. An entire topography of the human karyotype, both normal and pathological, began to be delineated. It began to seem that "apparently normal human phenotypes are occasionally compatible with gross rearrangements of chromosome structure" (Hungerford et al. 1969, 78)[16] and "ostensibly normal" people often had chromosomal anomalies. The body's clinical state could be understood as "ostensible" health,[17] as the anomalies of the chromosomes called the functioning body's legitimacy and soundness into question.

In his address to the 1966 Chicago conference, Penrose, then president of the Third International Congress of Human Genetics, characterized human genetics as a convergence of cytologists, geneticists, biologists, pathologists, and physicians. This interdisciplinarity made standardized nomenclature particularly important, for these different groups "tend to misuse each other's favorite expressions." As techniques for chromosome recognition improved, he said, much material collected now would be "almost useless to future investigations a long way ahead."[18] Standardization had both disciplinary and historical elements. It could permit knowledge to move from a cytogeneticist to a pathologist and from a scientist in one period to another scientist twenty years later. Standardization made present knowledge useful to future investigators.

Descriptions of the beauty and elegance of the human chromosomes are common in this literature, and Penrose, in his 1966 address, invoked the aesthetic appeal of the karyotype. "It is now just ten years since the veil of mystery was drawn aside sufficiently for the beauty of the human karyotype to be appreciated." In those early days, he suggested, the potential of the karyotype to explain human disease had not been sufficiently appreciated. It had been "impossible to foresee the wealth of variation which would soon be discovered. Almost every day some new aberration is seen." Deletions, duplications, inversions, reciprocal translocations, ring chromosomes, trisomies, singles and triploids and even tetraploids had been identified in six short years of intensive research. There was also "an incredible array of mosaics," and still more abnormalities remained to be detected. "It is easy to get carried away by the detectable peculiarities and to forget that much underlying variability is still hidden from view until some new technical devise discloses the finer structure of chromosomes."[19] Referring directly to standardization, Penrose proposed that nomenclature should be kept "sufficiently fluid" to allow for "essential changes in the future."

Human cytogenetics was in fact about to undergo a sea change, with the introduction, in 1968, of staining techniques that resolved entirely the questions of chromosomal identity. After 1970, chromosomes were increasingly identified by staining techniques rather than by measurement, replication rates, centromeric index, size, shape, or the presence of satellites (Sumner 1982). This change made reference books obsolete, resolved old debates, and, as might be expected, raised entirely new questions about chromosomal architecture. Quinacrine dihydrochloride, or Q-banding, described in a 1968 paper by Caspersson and colleagues, was the first "modern banding technique." Exposing prepared chromosomes to quinacrine produced characteristic bands of different fluorescence, and these bands were different on different chromosomes and in different species. It was not entirely clear why quinacrine produced this effect, but it was nonetheless a useful effect for cytogenetics.[20] Other banding stains came into laboratory use in the 1970s, including Giemsa banding (named after a nineteenth-century chemist) and C-banding. Earlier programs for visually identifying the human chromosomes were no longer of so much practical importance when chromosomes could be easily differentiated and matched up by their staining patterns, and although the numbering system was preserved, it ceased to be a subject of professional debate.

The Paris conference in 1971, the fourth conference on standardization, took up the question of banding patterns and began to systematically address a broad range of new "chores" relevant to human cytogenetics (Paris Conference 1971). For the modern cytogeneticist, the rules established at that conference may be far more relevant to everyday practice than those de-

MOMENTS OF TRUTH IN GENETIC MEDICINE

veloped at the Denver conference of 1960. Yet the Denver conference had established a pattern of interdisciplinary and international cooperation in naming and numbering that greatly facilitated the development and growth of human cytogenetics and that became a model for later standardization initiatives.

From my perspective, the standardization of the human chromosomes in the 1960s provides a compelling historical case study precisely because of the ambiguity of the chromosomes. They did not always make their identities clear, and naming and classifying them involved delicate social and technical negotiation—squashing the slide with the proper tempo and pressure and making sure that Patau was invited to the second conference, for example. Later, such negotiations shifted to other things, to names for genetic diseases, translocations, cell lines, and so on. The process, however, has remained a critical part of human cytogenetics and human genetics in general. A massive technical literature is devoted to naming things, and such labor provides insight into the structure and organization of modern biomedical science.

CHAPTER 5

[TWO PEAS IN A POD]

Twin Science and the Rise of Human Behavior Genetics

Twin studies require resolution of the question of zygosity—whether same-sex twins are derived from a single fertilized egg or from two different fertilizations. The "twin method" compares these two types of twins, and knowing which twins fall into which category is crucial. Scientists in the immediate postwar period determined zygosity through blood typing, anthropometrics, fingerprint analysis, and comparison of obvious phenotypic qualities such as hair color, eye color, and facial features. But in 1961 a widely cited Swedish study compared the accuracy of these technical, metrological means of diagnosis with the accuracy resulting from asking the twins themselves a single question: "Were you and your twin as alike as two peas in a pod?" By this group's calculation, of those twin pairs in which both twins answered yes, 98 percent were in fact monozygotic (identical). In other words, asking the twins themselves this particular question was as reliable in the determination of zygosity as were blood testing, phenotype analysis, and fingerprinting combined. Twins, researchers found, generally knew something about their own cellular origins (Cederlof et al. 1961).

The paper by Cederlof and colleagues, and papers by several other groups replicating its results, are still cited as justifications for using questionnaires to determine zygosity in large-scale populations of twins. In fact, large-scale studies of twins would be impractical without this inexpensive strategy, the "questionnaire method of determining zygosity." From my perspective this method fuses social and technical knowledge, or folk and scientific knowledge. An irretrievable gestational experience—having been derived from a single egg—is inferred on the basis of social consensus, and technical analysis testifies to the legitimacy of social perception: people know whether twins are monozygous or dizygous, and their knowledge has become a convenient surrogate for the results of the blood test, the cranial x-ray, examination of the birthing materials, or the DNA probe (Chen et al. 1999; Peeters et al. 1998; Goldsmith 1991).[1]

Twins are an important resource for research in human behavior genetics. Their knowledge of their own bodies has been embedded in the technical knowledge produced about behavior and genes. Folk or social knowledge has also been invoked to justify the behavior genetics research program. Behavior geneticists in the 1960s and 1970s commonly constructed their work as the mathematization of everyday experience. By demonstrating the genetic causes of human behavior, they suggested, they were validating what was already widely known by people in general: that talents, intelligence, and personality were inborn.

In this chapter, I explore twins as physical and social resources for the explosion in human behavior genetics after about 1955. I emphasize the practices involved in the creation and use of vast population registries of twins. I look particularly at the National Academy of Sciences–National Research Council Veteran Twin Registry, developed after 1955 and eventually including almost sixteen thousand pairs of white male twins who had served in the armed forces in World War II. My chapter title, drawn from the "questionnaire method of determining zygosity" used in the Veteran Twin Registry and in many others, is suggestive of my central themes. The image of "two peas in a pod" resonates with Mendel's peas and the critical role of peas in the history of genetics; with the agricultural origins of the eugenics movements around the world, which drew on the commonsense knowledge of farmers to make claims about human society; and with Darwin's profound debt to folk knowledge of domestic breeding in the development of his theory of evolution by natural selection. I suspect that the "peas in a pod" image was simply a convenient and familiar way to express biological sameness, using a commonplace (Western) phrase that subjects would readily comprehend. But it is also a remarkably rich image, a cultural shorthand for both folk knowledge and the social meanings attached to things and people that are "the same."

By the late nineteenth century, heredity had long been understood to be involved in what might be called "instinctual" behaviors such as maternal love. The British eugenicist and statistician Francis Galton began promoting twins as resources for behavior genetics in 1875, a date that stands as the traditional starting point for both the "twin method" and the modern nature-nurture controversy. Galton did not propose the modern twin method, but he is commonly credited with having done so (Rende, Plomin, and Vandenberg 1990). At the same time, Galton's cousin Charles Darwin saw some human behaviors as adaptive, which was clearly based on the assumption that they were hereditary, and both Darwin and Galton were presumably drawing on a commonsense or folk knowledge of "human nature."

Later studies of animal behavior drew on the newly standardized rats and mice of the 1920s and 1930s. The largess of the Rockefeller Foundation supported scientific research into animal behavior, expecting such data to provide a guide to building a better society. When Alan Gregg, as director of the Rockefeller Foundation's Division of Medical Sciences, announced that "wise matings" were more important than "$800,000 high schools," he was expressing the social conviction of many American elites that intelligence, a highly valued form of cultural capital, was hereditary (Paul 1991). During the same period, eugenicists developed baroque narratives of genetically driven human behavior. Some of their hypotheses retain a certain bold charm, for example their beautifully transparent collapse of the biological and the social via the supposedly sex-linked gene for thalassophilia, or love of the sea (G. E. Allen 2001). By the 1930s, some geneticists, psychiatrists, and psychologists in the United States and elsewhere had begun serious research on the genetic origins of human behaviors, including mental illness. Some of this research involved the use of twins, generally tracked through institutional populations. By 1945, then, genes for behavior, as manifest in twins and other human groups and in animals, had been the focus of scientific and political interest for sixty years.

Only in the postwar period, however, did the infrastructure of professional human behavior genetics develop. New journals, societies, faculty appointments, and international conferences focused on human behavior genetics in the 1960s, the decade widely regarded by participants as the

MOMENTS OF TRUTH IN GENETIC MEDICINE

watershed for the field. The first textbook with a strong emphasis on human behavior genetics, called simply *Behavior Genetics*, was published in 1960 (Fuller and Thompson 1960), the canonical moment of transition for later participant-historians in the field. By the late 1960s, programmatic statements about the new and improved status of human behavior genetics were common in the international journal of twin research, *Acta Geneticae Medicae et Gemellologiae*, which was started in 1951 by the geneticist Luigi Gedda, at the Gregor Mendel Institute in Rome. In 1970 the journal *Behavior Genetics* began its first rocky year of publication, the first issue featuring chromosomes in mitosis on the front cover and containing eight papers, six on mouse behavior, one on twins and schizophrenia (a classic topic in the field), and one on personality in Turner syndrome. The opening commentary from the editors proclaimed behavior genetics a "fertile hybrid" (Vandenberg and DeFries 1970). Around the same time, the Behavior Genetics Association acquired the formal qualities of a professional society, with a president, prizes, annual meetings, and membership fees.[2] As J. N. Spuhler put it in 1967, "no well ordered body of knowledge that deserved to be called 'human behavior genetics' existed until well after World War II" (v). One of the most important forms of human capital for this new enterprise in the 1960s was the twin.

By my rough calculation, based on published data about twin registries, from 1950 to 1970 approximately forty thousand twin pairs, identical and fraternal, were enrolled in medical studies of a wide range of behaviors and diseases in Europe and the United States. In 1966, the World Health Organization reported that it had surveyed thirty-nine such twin registries, ranging in numbers of twins from forty to twenty thousand pairs. Such databases did not exist before 1945. Twin researchers had "collected" twins they encountered in institutional or clinical settings, working with such "selected" twins in relatively small numbers. But in the 1950s and 1960s, for many different institutions and investigators in many different national and local settings, creating a population-wide, unbiased, and unselected database of identical and fraternal twins seemed a highly worthwhile endeavor (Hauge et al. 1968; Essen-Möller 1970).

What did those who built the massive twin registries of the 1960s believe they were building? Twins were specifically understood to be research tools suited to the study of complicated, vague genetic causation rather than classic, dramatic genetic diseases. The classic genetic diseases such as hemophilia or Huntington disease could be embedded in family pedigrees that il-luminated their mode of transmission. But complicated or ambiguous be-haviors and chronic or poorly defined diseases were much more difficult to understand in genetic terms merely by looking at a family pedigree. In many cases the conditions did not even "run in families" in the traditional sense. More than 90 percent of people with schizophrenia have nonschizophrenic parents, for example. Twins were meant to answer questions about the ge-netic origins of traits and behaviors that were not necessarily genetic. Their bodies were subtle, sensitive reflections of the actions of genes through which difficult-to-classify bodily states could be studied. For studies of human behavior, they were the ideal organism.

Twins are a node in a thick network of meaning, technology, and social organization. They are a window, a point of entry, a historiographical ex-perimental organism, if you will. Just as behavior geneticists have used twins to tell stories about heredity, I am using twins to tell a story about the historical development of technical knowledge.

Like many other observers of behavior genetics, I am reasonably skepti-cal about the overall project and reasonably certain that human behavior is powerfully shaped by culture. The varieties of human society in history and around the globe suggest at least that the species is extremely flexible. But in my exploration of twin research, I am not particularly concerned to refute the scientific findings. My focus is on behavior genetics as an enterprise that provides a signal example of the expansive, fungible quality of genetic ex-planations.

Twins and Genetics

Francis Galton is usually considered the first scientist to pay serious atten-tion to twins as potential research organisms. It is important to note that, whatever his ultimate intentions, the eugenicist Galton framed his project as a test of the hypothesis that the environment shapes personality more than heredity does. His first reference to twins, in his 1874 book *English Men of Science*, proposed that identical twins became unlike under the "gradually accumulated differences of nurture" (quoted in Rende, Plomin, and Van-denberg 1990, 278), and many of his later references to twins reiterate a similar point. Galton's method was to compare twins through time, on the theory that twins who began life very much alike might change as they had different experiences. He also postulated that twins who were quite dissimilar at birth might become more alike if they lived in the same environment (Galton 1875).

Galton's data-collection methods were casual. He asked friends and acquaintances if they knew any twins, and then he collected information about how these twins were different and how they were similar. In his 1883 *Inquiries into Human Faculty and its Development* (cited in Newman, Freeman, and Holzinger 1937, 4–5), he reported obtaining information on thirty-five pairs of twins who were considered by their friends and families to be difficult to distinguish. He also collected information, via correspondence, about twenty pairs of twins who were thought by those who knew them to be no more alike than siblings generally. From this sample of fifty-five twin pairs, Galton concluded that twins seen by friends and family as very much alike shared many physical ailments or anomalies and seemed to have similar dispositions. He proposed that when these twins had differences in disposition, it was the result of illness or accident. Consistent with this hypothesis, Galton also found that twins who were perceived as unlike by family and friends did not tend to become more alike even if they lived in the same environment all their lives. Galton concluded that "environmental differences, such as are to be found in the same community and at the same time, produce slight change in the individual's physical and mental makeup" and that "physical and mental characteristics are determined chiefly by inborn nature" (5).

The modern twin method often attributed to Galton was in fact proposed in the 1920s, in Germany. Twin research in Europe, and particularly in Germany, was generally more hereditarian than research in the United States. Twins seem to have been interesting primarily as models for genetic determination, and out of this reductionistic perspective came one of the most important ideas in twin research, namely, that twins should be compared not to their own family members but, statistically, to other twins for concordance ratios. A form of the "twin method" appeared in 1924 in a paper by the German dermatologist Hermann Siemens (Rende, Plomin, and Vandenberg 1990).[3] Unfortunately Siemens and his students believed that any trait for which identical twins differed in any way could not possibly be genetic; as a corollary, any trait they shared had to be genetic. This had the advantage of greatly simplifying the nature-nurture question, but as contemporary critics pointed out, height might well differ in identical twins (it often did) because of nutrition, accident, or illness, while being nonetheless genetically mediated.

Mental illness was the primary focus of much of this early German twin research. Schizophrenia, a disabling mental illness affecting approximately 1 percent of the world's population, characterized by delusions, hallucinations, and formal thought disorders, was one of the most widely studied mental illnesses. The diagnostic category of schizophrenia was first conceptualized by the Swiss psychiatrist Eugen Bleuler in 1911. Bleuler believed that schizophrenia was organic and that "heredity does play a role in the etiology of schizophrenia, but the extent and kind of its influence cannot as yet be stated" (quoted in I. Gottesman and Shields 1972, 16). Ernst Rudin, who had studied with Bleuler in Zurich, in 1916 undertook studies of schizophrenia as a Mendelian disorder, assessing 701 sibships and finding correlation frequencies that suggested schizophrenia was caused by two recessive gene pairs (cited in I. Gottesman and Shields 1972, 20; see also Rosanoff, Handy, and Rosanoff 1934). German eugenicists such as Rudin, Emil Kraepelin, and Alfred Ploetz (Weindling 1989; S. F. Weiss 1987; Muller-Hill 1998) were active participants in this twin research. Later, this legacy of psychiatric genetics—its historical origins in a culture that produced the Holocaust—may have contributed to scientific skepticism in the United States about any hereditary explanation for mental illness (I. Gottesman and Bertelsen 1996).[4]

Probably the largest group of twins brought together explicitly for behavioral studies before 1950 were the nearly four thousand twins catalogued by Franz Kallmann, a psychiatrist who devoted himself to the study of schizophrenia. Because he was half Jewish, Kallmann was removed from his position at the Kaiser Wilhelm Institute in Berlin in 1935. He emigrated to the United States in 1936, taking a position at the New York State Psychiatric Institute (Muller-Hill 1998, 122). In New York, Kallmann began working with twenty psychiatric institutions with a combined population of seventy-three thousand people and twelve thousand new admissions yearly. Such a population included many persons diagnosed with schizophrenia and many persons who were twins. He found very high rates of concordance, up to 86 percent by some calculations, in identical twins (Kallmann 1938, 1946; Slater 1953, 18–22). Later studies of schizophrenia in Finland, Norway, Denmark, and the United States found much lower concordance rates (6 to 48 percent) than Kallmann's studies. Kallmann has the dubious distinction of proposing, in 1935, the involuntary sterilization of family members of schizophrenics on the premise that they were carriers of the disease, a proposal that even the Nazi racial hygienist Fritz Lenz found problematic. Kallmann was also convinced homosexuality was genetic, and only very reluctantly abandoned his conviction that suicide was genetic, finally deciding after many years of suicide research that suicide was "so complexly determined, with genetic factors playing such a minor role, that the suicide of each of a pair of twins would be largely coincidental and therefore in any case very rare" (Kallmann et al. 1949, quoted in Slater 1953, 21).[5] Because of these views and research programs, Kallmann was a controversial figure who could sometimes be invoked to call into question all research in behavior genetics.

In the United States, one of the most important early twin studies was carried out by the biologist Horatio Newman, psychologist Frank N. Freeman, and statistician Karl J. Holzinger, all at the University of Chicago. Their major study of twins began in 1927 and was completed and published ten years later.[6] They worked with only 119 pairs of twins—fifty monozygotic (identical) and fifty dizygotic (fraternal) reared together and nineteen monozygotic twins reared apart. But their study was extraordinary in its depth and complexity, and the resulting text is widely viewed as one of the classics in the field.

Identical twins separated at birth were, of course, the ideal twin pair. In 1925 the *Drosophila* geneticist H. J. Muller, who later won the Nobel Prize for his studies of radiation mutagenesis, published the first scientific report of such a pair (Muller 1925). The circumstances of identical birth and early separation may seem extremely unlikely, but in the last seventy-five years almost two hundred such pairs have been identified and studied.[7] As the clinical narratives recorded by Newman, Freeman, and Holzinger suggest, many circumstances could bring such twins together. Six of their twins "had never heard of each other's existence until they were brought together by some curious freak of circumstances" (143). These included chance encounters in cities distant from the twins' homes, an acquaintance of one rushing up to greet the other, a birth certificate encountered while cleaning an attic, and a nun who had taught one twin finding herself sitting next to the other on a bus trip.[8] Most twins, of course, are reared together, and most research has involved twins who could be presumed to have experienced the same, or very similar, environments.

Some thirty years before the "two peas in a pod" program, Newman, Freeman, and Holzinger adopted a similar approach. They were asking the identical twins separated at birth to travel to Chicago to be tested and compared for their research, but "because of the great expense involved in bringing these separated twins to Chicago, no chances were taken that any of them might prove to be fraternal twins." In order to avoid paying to bring together twins who were not identical, they asked the twins, "are you or have you been at some time so strikingly similar that even your friends or relatives have confused you?" Although it is unclear (to me at least) how twins separated at birth could reasonably answer such a question, Newman and colleagues said that the answer was highly reliable, providing "proof of the monozygosity of the separated pairs" (135).

Their other twin group consisted of twins "found chiefly by inquiry in schools in Chicago or its suburbs." They were not selected for illness or chosen because they were similar. Although the twins thus collected did not constitute a complete population of all twins born in a given geographical region, they were more representative of twins in general than were those collected by tracking institutionalized populations (28).

The twins who came to Newman and his colleagues' laboratory at the University of Chicago were subjected to two days of testing and observation. They were compared for physical appearance (height, weight, hair color, handedness, skin texture, palm prints and fingerprints, birthmarks, etc.) and performance on the Stanford-Binet test, the Otis Self-Administering Tests of Mental Ability, the Stanford Achievement Test, and the Downey Will-Temperament Test (29). The team and their graduate students also collected information on school performance from teachers, on interests from the twins themselves, and on behavior from parents. People who knew the twins were asked to describe their dispositions, ambitions, talents, and defects.

In other words, the twins gave up two days of their lives (perhaps more), endured physical examinations and many standardized tests, and permitted the scientists to invade their social worlds to do interviews in which their flaws of temperament or performance might be discussed and recorded. I would suggest that it is important simply to notice the presumed shared values or shared assumptions permitting such research, particularly in the light of the complexity of gaining cooperation from human subjects in the twenty-first century. The twins were not just data points. They were actively enrolled in the project, and whether this active enrollment reflected their love of scientific truth making or the hegemonic power of technoscience to co-opt all cultural belief systems, or whether the twins and their families were explicitly bullied or implicitly intimidated by the high status of the research group

MOMENTS OF TRUTH IN GENETIC MEDICINE

Twins profiled by Newman, Freeman, and Holziger. From Newman, et al.,
Twins: A Study of Heredity and Environment, 319. Published with the
permission of the University of Chicago Press.

at the University of Chicago, is unresolvable without further research. What
is clear is that they participated—for whatever reasons—in a relatively unpleasant endeavor that drew on their life histories to tell stories about natural truth. Their experiences are part of that knowledge. Indeed, one could
even argue that they are the important part.

Newman, Freeman, and Holzinger conceived of their project as "holding
constant" either heredity or environment. For the fraternal twins reared to-

gether, environment was held steady while heredity differed. For the identical twins reared together, both heredity and environment were presumably the same, yet the two individuals often differed from each other in personality and vigor, a phenomenon the research team attributed to slight variations in fetal environment. Finally, Newman and colleagues compared identical twins reared apart, an experiment in which heredity was held constant and environment differed.

They found that "in most of the traits measured the identical twins are much more alike than the fraternal twins . . . this is true of physical dimensions, of intelligence, and of educational achievement. The only group of traits in which identical twins are not much more alike consists of those commonly classed under the head of personality." Indeed, for many of the personality traits, the "correlations of identical twins are but little higher than those of fraternal twins." This tendency of identical twins to differ in personality was later described as the phenomenon of "twinning," the social process through which identical twins differentiate from each other and fill unique niches in the economy of family life. Newman, Freeman, and Holzinger had spent a lot of time thinking carefully about twins, and in their concluding chapter they proposed something very like the modern theory: "Each twin is part of the environment of the other, and a part which, while it does not differ much in original nature, comes to differ more because of the differentiation of attitude and behavior growing out of their mutual association" (353–55). This early, detailed, and highly sympathetic study of twins suggests how much could be learned, by investigators willing to be self-critical, from the simple natural experiment of multiple birth.

The first systematic effort to create what might be called a twin registry was undertaken in Germany, by the German psychiatrist H. Luxenburger (Luxenburger 1930, cited in Slater 1953).[9] Luxenburger used hospital admissions for a wide range of psychiatric disorders, drawing on all hospital records in a given state or region. The names collected in this way were then sifted by place of birth, and the registry offices for these regions were asked to make a search to discover whether the patient had been born a twin. If the patient had been a twin, Luxenburger sought further records of both twins to establish discordance or concordance for the mental condition that led to hospitalization of one twin. In other words, he explicitly sought to track the entire population. In its rough way, this effort suggested the inadequacies of earlier, single case report methods. Luxenburger found that his

　　　　　　　MOMENTS OF TRUTH IN GENETIC MEDICINE

twins were about 80 percent discordant for the psychiatric disorders in question. Single cases reported in the literature up to that time had been about 80 percent concordant. Experts looking at twins apparently only noticed, or only published papers, when the twins were concordant for mental illness. Perhaps proving the value of Luxenburger's approach, around the same time, also in Germany, Johannes Lange was collecting data on twins from the Institute for Criminal Biology, Bavarian prisons, and the Genealogical Department of the German Institute of Psychiatry in a carefully selected series intended to show (and supposedly showing) that criminality was hereditary (S. F. Weiss 1987; Slater 1953, 8–9).

What might count as a twin registry is, of course, a matter of judgment. Some collections of twins brought together for a single study could be seen as registries by some authors but not by others. In a 1968 report by Hauge and co-workers at the Institute of Medical Genetics, University of Copenhagen, for example, Essen-Möller's catalogue of twins with psychiatric diagnoses collected in the 1940s was interpreted as an early registry, whereas Franz Kallmann's far larger collection was not. By 1970 Essen-Möller reported twenty thousand twin pairs in the registry at the Psychiatric Department of the University of Lund. Hauge and colleagues construed the Danish Twin Registry, begun in 1954, as the first population-based registry. It began with a study of official birth records for 1870–1910 and kept records on all twins born in these years, including those already dead by the time the registry was created, something not always done in twin registries but apparently reflecting the idea that twin births and deaths could be used in some kinds of epidemiological work. The number of twin pairs born in Denmark during this period was 37,914, but more than 64 percent of these individuals died before the age of six.

By the late 1940s and early 1950s, a twin registry that was a complete population screen had become the methodological ideal. These new twin registries were expected to have the attributes most sought-after for scientific research: unselected, complete, with clear ascertainment of zygosity, and with a governing body—a committee or institute—to collect data and blood and manage access. The twins thus catalogued would be a ready-made research population, almost a standardized organism like the laboratory mouse, to be taken off the shelf if one were studying alcoholism, graying hair, earnings potential, cancer, or cardiovascular disease. They were a prepackaged set of special bodies, aligned through a certification system of zygosity diagnosis, mental testing, and physical examination. And they were biological products linked by the specific historical experience of the shared womb.

Useful Veterans

One of the largest twin registries created in the United States at this time was the National Academy of Sciences–National Research Council (NAS-NRC) Veteran Twin Registry. These sixteen thousand white male veteran twins were brought together in a database managed and funded by the NAS-NRC and the Veterans Administration (VA), which drew on the records of the Federal Bureau of Investigation, credit organizations, and the Internal Revenue Service. The resulting registry was expected to be "worth millions" by one adviser to the NAS.[10] It was certainly one of the most ambitious registry projects and one of the few that focused on adult twins, adults usually being harder to track than children.[11]

The veteran twins were part of a much larger program to use the veteran population for medical research, the Medical Follow-Up Agency, which still exists today. In one 1953 report on the medical follow-up of all veterans (not just twins), an NRC committee noted that the veterans should be tracked for several reasons: the population of veterans was very large ("quite large enough to be forgiven the bias as to age, sex, and health at entry"); veterans had endured extreme stress, trauma, and unusual disease; and the population was documented in the central index by means of punch cards, through the VA medical care program, with its "low threshold for recognition of a disease state," and through the numerous ancillary records, the central file of the Armed Forces Institute of Pathology, and the central locator files. In other words, the veteran was written up, embedded, textualized, and therefore an ideal object for medical observation.[12]

This ambitious epidemiological program began in 1946, a few months after Colonel Michael E. DeBakey, of the Army's Office of the Surgeon General, wrote a memo proposing that the war had created "an enormous amount of material of great clinical value" (Berkowitz and Santangelo 1999, 3). DeBakey, one of the creators of the MASH (Mobile Army Surgery Hospital) units in the Korean War and later famous for his innovations in cardiovascular surgery, proposed that a long-term follow-up study of veterans could provide information about disease and health that was unavailable for any other population. Surgeon General Norman Kirk was convinced, a com-

MOMENTS OF TRUTH IN GENETIC MEDICINE

mittee was created at the National Research Council, and DeBakey, still in uniform, was assigned to work on the project at the NRC with the NRC epidemiologist Gilbert Beebe. An enthusiastic group of medical and military leaders endorsed this effort to "follow a whole generation of men and trace their life history" (Berkowitz and Santangelo 1999, 4). The VA would pay for the vast multipurpose study ($850,000 in 1947 alone), and the Institute of Medicine of the NAS-NRC would manage it. Any interested scientist or physician could apply to the NAS-NRC to use the database. The veterans were a resource with many possible meanings and uses.

Beebe favored the idea of "mass statistical studies" rather than the follow-up of individual soldiers, partly because such statistical studies could be done entirely from medical records, without any need to examine or speak to the veterans. It was epidemiology by database, more or less. So, for example, Beebe favored studies of death rates from individual diseases over time. DeBakey promoted a more clinical approach in which the outcomes of therapy in aggregate cases or the outcomes of particular injuries could be compared.

Studies drawing on both approaches were eventually carried out. As early as 1948 the Committee on Veterans Medical Problems and the Follow-Up Agency were supporting medical studies of veterans in thirty different centers around the nation. These included studies of psychoneurosis, peripheral nerve injury, vascular injuries, and infectious hepatitis (Berkowitz and Santangelo 1999, 4, 13).[13] Later work included studies of mortality and psychiatric problems in former prisoners of war, the value of x-ray screening for the detection of tuberculosis at induction, exposure to microwave radiation, and the long-term consequences of head injury, including epilepsy. Former prisoners of war could give lessons about "the human biology of extreme stress" and "levels of adjustment in the main areas of life, in the years since repatriation." And veterans could provide insight into multiple sclerosis, cardiovascular disease, alcoholism, the viral origins of cancer, the long-term effects of amputation, and "body build and mortality." Corregidor became a "risk factor" in psychiatric studies; and still later, exposure to radar equipment during World War II was used to assess the risks of cellular phone use in the 1990s.[14]

Veterans were thus enrolled in a massive cold-war project that construed them as unique biomedical research subjects. Serving country, medicine, science, and Wall Street, they placed their testimony and their body parts in the circulating networks of politics, nature, and commerce. The organizations tracking veteran experiences included the Armed Forces Institute of Pathology, the Walter Reed Army Institute of Research, the U.S. Public Health Service, and the National Institute of Mental Health. Veterans with Hodgkin disease could be identified through the Lymphatic Tumor Registry of the Armed Forces Institute of Pathology, and these records could be overlaid with those of the Surgeon General, the National Archives Records of Admissions to Red Cross Hospitals, the American Medical Association, the Retail Credit Company, and "one agency which prefers to remain anonymous."[15]

In breathtaking detail, the structure of surveillance was laid out. Foucault's perspective on the control of the body in the modern state was fully manifest in the Medical Follow-Up Agency, which was an alliance of military medicine, the NAS, Congress, and academe. The more than five hundred publications growing out of this research constitute an inscriptive empire of data, numbers, facts, and conclusions drawn from human beings with the unusual historical qualities of having been wounded at Iwo Jima, held in a Nazi prison, exposed to pathogens or risky agents, or simply pulled into the record-keeping nexus of the American armed forces (Berkowitz and Santangelo 1999, 99).

And yet, despite all these records, all this voluminous textuality, despite this obsessive tracking of bodies and diseases in veterans, the subterranean quality of having been born a twin was not immediately visible. The fetal experience could not be accessed by looking at induction records, and the moment of birth, when two bodies emerged from the same womb, was not documented in the records of the U.S. Army. The intimate fact of having come into the world with another person was not something that could be seen in the medical records of young soldiers. Finding twins, then, would require an entirely different overlay of text and surveillance.

Finding Veteran Twins

Early in the planning stages for the Medical Follow-Up Agency, in 1946, NRC statistician Beebe proposed the possibility of tracking veteran twins, but the proposal was quickly abandoned when it became clear that twin status would be neither obvious from the existing records nor easy to determine from some other record such as the census.[16] The idea was resuscitated by 1955, and an initial search of the central alphabetical files of military

MOMENTS OF TRUTH IN GENETIC MEDICINE

personnel turned up a few pairs of twins, but "the method seemed expensive."[17] Yet interest in compiling twins was high. The idea appeared and reappeared several times in the early history of the Medical Follow-Up Agency. Why?

Certainly one reason was that twins were expected to be uniquely useful in the identification of "hereditary and environmental factors in disease." A 1957 summary document, in which research on veterans was described, stated that serious planning for a twin registry had begun, and "comparative studies of identical and fraternal twins" would be a "resource for special studies by human geneticists and others."[18] The report went on to suggest that twin studies would "eventually prove to be our most significant undertaking."[19] By this time, the Medical Follow-Up Agency was providing researchers with access to human populations for a wide range of studies, with funding from many government agencies and private foundations. And the twins were not even collected yet. The NRC was still struggling to figure out how to find them and to reliably diagnose zygosity. Why would the twins be the agency's "most significant undertaking"? Perhaps because they were resources for the study of diseases such as cancer, cardiovascular disease, and other major, chronic conditions that were expected to emerge as key issues in public health in the late twentieth century.

The human geneticist James V. Neel of the University of Michigan drafted a justification for the twin project in June 1956. In that proposal, Neel said the twins were valuable because they could be used to develop a better understanding of diseases that were very different from the kind generally understood to be "genetic diseases." "Advances have been made in the understanding of human genetics in recent years," Neel wrote, "but these have been largely dependent on the recognition of characteristics more or less sharply determined by particular genes." The really important questions in public health, epidemiology, and clinical medicine, he suggested, related to much more complicated disease states, to conditions unlikely to be explained by the "identification of specific genes." These conditions, such as cancer and heart disease, presumably had some genetic component, he suggested. They were "likely to depend on a multiplicity of genetic factors for which, in man, there is presently little hope of identifying the components, or differentiating their interactions and their distribution in the population or in lines of descent."[20] Twins, however, could make these genes visible to scientists.

Twins were significant and important biomedical resources, then, because they could support genetic studies of complex disease states in ways that could account for hereditary difference. Sharply determined genetic diseases were not as critical for clinical medicine and public health, but cancer, cardiovascular disease, and other major, chronic degenerative diseases that might involve a "multiplicity of genetic factors" posed problems directly relevant to everyday medical care. "It is likely that the major contributions to an understanding of the genetic basis of those diseases which are now at the forefront of medical interest will be realized in the continued study of this roster over future years." Twins were the biological keys to this "broader aspect of human genetics," which could be "profitably pursued by comparisons of the medical histories of one-egg and two-egg twins."[21] Twins could help human genetics move into public health.

Neel was one of the leading figures in medical and human genetics at this time. He was in the midst of establishing a high-profile, well-funded program in human genetics at Ann Arbor; he was an international spokesman on the risks posed by exposure of human populations to mutagens such as radiation, the chief geneticist for studies of the genetic effects of radiation on the atomic bomb survivors, and an experienced practitioner of human population genetics and genetic epidemiology (Lindee 1994; Neel 1994). Despite his support for the Veteran Twin Registry, he was a bit skeptical about the twin method and its usefulness. In 1957 he proposed that virtually all work on twins and genetics up to that time had been deeply flawed, either because genetic claims had been accepted uncritically or because zygosity had not been carefully determined. "Twin studies have been seriously handicapped," he said, partly because identical twins who grew up together inherently confounded nature and nurture, and also because no studies yet had been able to track large numbers of twins who were homogeneous for age, race, sex and socioeconomic status. While the assessment was moderately unfair—some twin research by 1957 had paid careful attention to both environmental influences and zygosity—the claim had the advantage of justifying the Veteran Twin Registry. Large numbers of twins, homogeneous as to age, race, sex, and socioeconomic status, were precisely what the registry might offer.[22] The only problem was how to find these homogeneous, administratively embedded twins.

The solution was the birth record. In 1955, a plan to find veteran twins by tracking down multiple births through state offices of vital statistics was tested with the help of the state of Maryland. The protocol depended on the mathematico-historical fact that the vast majority of men who served in World War II were born between 1917 and 1927. In an act of statistical faith, Beebe asked the Maryland office to find all birth records of white male twins, both born alive, in these years. The names of these twins were then matched against Veterans Administration, Navy, and Army central files, and "it was found that 30 percent of the individuals had survived and entered service, and 16 percent of the twin pairs had jointly survived and entered service." This successful search was also relatively cost-effective: each usable pair of twins had cost about $10 to find.[23]

In order to create the larger roster, the Follow-Up Agency would have to repeat this exercise on a grand scale, asking vital statistics offices in the United States to search their birth records for the years 1917–27 and provide photocopies or transcripts of records for all white male twins, both of whom were live-born. Beebe estimated that twenty-five million birth records could be searched and that this would yield, on the basis of the Maryland study, about fifty thousand pairs of live-born white male twins, which should, again on the basis of the Maryland study, yield about six thousand to ten thousand twin pairs who served in the armed forces. A roster of ten thousand unselected twin pairs would be the largest in the world up to that time.

The NRC had surveyed "every state and independent city health department in the country" to find out whether such a search were possible, and thirty-eight states said it was. The letter sent to state registrars in the summer of 1956 asked these state employees to help in the creation of a database that would shed light on the genetic factors in human disease. State registrars were told that twins offered a "unique opportunity" to evaluate heredity and environment. Twin studies in the past had "fallen far short" primarily because of the "great difficulty of bringing under continuing long-term observation a sufficient number of pairs of twins, reasonably homogeneous with respect to certain important variables." Veterans might provide these large numbers of homogeneous twins, but "unfortunately twins of whom both members entered service are not identified as to their relationship in military and veterans' records." It would therefore be necessary to start with the birth records of male twins and match these records with VA records. NRC statistician Bernard Cohen told state registrars that the initial pilot search of 24,763 birth certificates in Maryland took 17.5 hours, a rate of fourteen hundred records per hour, and yielded fifty male pairs. "I would greatly appreciate your letting me know if such an undertaking is possible in your office, and if it is, what your estimates are of the time that would be required and the costs."[24]

Thirty-nine states cooperated, and over the next few years the birth records of 54,000 twin pairs were matched with the records of the armed services. One system of surveillance interacted with another to produce a biological product: twins. For 23,000 of these pairs, neither twin was a veteran, and for 15,000 twins only one twin was a veteran, but there were 16,000 pairs in which both twins had served (Jablon et al. 1967, 137). With the facts of twin birth and military service more or less established, the agency now faced the equally daunting task of determining zygosity.

Zygosity

Without accurate zygosity diagnosis, the twins would be no more useful than any other population of white men of a certain age. They certainly could not be used to reach conclusions about behavior genetics if their own genetic relationships were unclear. Examining individually the more than thirty thousand people listed in the twin registry was not practical, and even the details recorded in the textual records of the VA, Federal bureau of Investigation (FBI), Internal Revenue Service, Public Health Service, and all the other organizations might not be enough to distinguish monozygotic from dizygotic twins.

The FBI had fingerprints for all inductees and had already agreed, in principle, to make these fingerprints available for the twin study. Although identical twins do not have identical fingerprints (fingerprints are developmental products), they do have very similar fingerprints, and fingerprints had been used to diagnose zygosity since the 1930s, always using a scoring method for similarity. But fingerprints were not considered the most reliable or easily interpreted form of biological marker.[25] The FBI also had physical descriptions of the inductees, though these, too, were not entirely diagnostic. Hair and eye color, height, and weight are suggestive but not conclusive guides to zygosity.[26]

Neel and Beebe, with the support of the Follow-Up Agency and the NRC committee, decided to create a subpanel of twins who had been physically examined, with blood samples, photographs, and body measurements, as a means of determining the best form of zygosity diagnosis. This subpanel was to be the most carefully documented set of veteran twins, the only group subjected to systematic examination in person, and the results from this

MOMENTS OF TRUTH IN GENETIC MEDICINE

panel would be used as a guide to the assessment of zygosity in all other twins. It was quite fortunate, therefore, that the most reliable method of determining zygosity turned out to be asking the twins themselves.

For the subpanel project, a list of 298 twin pairs was drawn up, in which both twins were alive, both had fingerprints on file with the FBI, and the last known address for both was in the lower peninsula of Michigan or in upstate Illinois. This latter requirement was because Neel was in Ann Arbor, and he would direct the medical students who would carry out the field work. A brief letter was sent to each twin, informing him that a study of twin veterans was to be undertaken, asking his cooperation, and promising him, at the completion of the study, a report on his blood types and "our opinion as to whether he was an identical or fraternal twin" (Jablon et al. 1967, 135).

Neel hired seven medical students to contact and arrange to see these 596 persons over the summer break.[27] Sometimes one or both twins could not be found. In all, 514 veterans (257 pairs) were questioned, examined, and asked to contribute blood, hair, and saliva samples to the laboratory. Each twin was classified by eye and hair color and interviewed about his life and occupational history. The blood samples were typed at the University of Michigan with reference to nine blood-typing systems, and the saliva was typed with reference to the ABH secretor trait. Sometimes a twin had some missing information, or twins were found to differ by only one test (casting doubt on the test), and so forth, and in the end only 232 pairs of twins were reliably diagnosed for zygosity in this subpanel.

Neel's doctors-in-training made it a point not to express any opinion about whether the twins were identical or fraternal during these field visits. When the field work was finished, all the participating twins were sent a letter stating that the Michigan group was ready to send out the results, but first, they would like the twins to express their own opinions. "It has been said that twins themselves know whether they are identical or not, and that their statements are more accurate than scientific tests. Before we send you our opinion, would you be so kind as to check your opinion on the enclosed post card and drop it in the mail box?" The results of this simple survey suggested that the twins were quite good at determining their own zygosity. They had been exposed over their lifetimes to "numerous lay and even professional opinions concerning zygosity" (Jablon et al. 1967), and those opinions were backed up by blood typing at a rate that permitted the blood analysis (messy, expensive, difficult) to drop away.

Meanwhile the fingerprints on record with the FBI, laboriously scored for similarity, proved to be significantly less reliable than the twins' own opinions.[28] As Jablon later calculated the results, the twins' own judgments of zygosity were contradicted by lab results only 4.3 percent of the time, whereas fingerprint and anthropometric (bodily measures) results were contradicted by the lab results (blood tests) more than 16 percent of the time.[29]

In their published paper, Jablon and colleagues went so far as to propose that when the twins disagreed with the technical experts, perhaps the twins were right. "More monozygous twins erred in the belief that they were dizygous (10.9 percent) than dizygous twins erred in the belief that they were monozygous (3.2 percent). This may reflect a mistaken belief that identical twins should demonstrate complete concordance in such traits as height and weight. However, we must recognize the possibility that some twin pairs who are concordant serologically and morphologically and classed by us as monozygous are nevertheless dizygous, and that the opinion of the twin pairs is as valid as our own" (Jablon et al. 1967, 150–51). This suggestion, published in the *American Journal of Human Genetics* in a paper announcing the availability of the registry, tentatively elevated the knowledge of twins above the knowledge of scientists.

The announcement framed the twins as a resource in the study of genetic factors in the diseases of middle and old age. The paper's authors acknowledged that the twins were presumably healthier than the population in general, with participation in the registry "restricted to twins both of whom survived to military age and both of whom were then found physically fit for military service." They might have added that both had also survived the slaughter of World War II, which made them not only relatively healthy but lucky. Although the panel was "neither representative nor cross-sectional," it had the important advantage of providing access to older twins who would soon be manifesting the chronic diseases of old age (Jablon et al. 1967). Once the registry was compiled, once the lists were made and the zygosity questions resolved, the "million dollars" worth of bodily resources would presumably start to pay off.

Using the Twins

As the twin registry was being planned in the late 1950s, the geneticist L. C. Dunn suggested "actively seeking the cooperation of the twins to be examined and, in a sense, making them and their interests part of the planning."[30] In practice, and for quite practical reasons, the interests of the twins came to loom large in the planning process. This was particularly true when the registry was completed and it became necessary to think about permitting researchers to exploit the new resource. Using the registry required balancing the ideals of good research with concerns for the privacy of the twins, the desire to avoid alienating or annoying the twins by repeated requests for participation, and the need to coordinate the efforts of competing research groups interested in the same questions. The NAS-NRC committee charged with overseeing the use of the twins negotiated these compromises with keen attention to the research subjects. The committee members' expectations of the reactions of the twins—what they would think and how they would feel—affected institutional reactions to virtually every proposal.[31] The imagined responses of the twins constituted a filter that shaped the research.

The first few possible research projects that would draw on the registry were discussed at a meeting in 1967, just after publication of the journal article describing the creation of the registry. The members of the Committee on Epidemiology and Veterans Follow-Up Studies included Brian McMahan of the Harvard University School of Public Health, Thomas Chalmers of the Tufts University Department of Medicine, DeBakey, then of Baylor University, and Neel, of the University of Michigan. They were joined by three VA representatives, Manning Feinleib of the National Institutes of Health (National Heart Institute), two Atomic Energy Commission physicians, and five staff members from the Division of Medical Sciences of the NAS-NRC. Those negotiating the uses of the twins, in other words, occupied a range of institutional and disciplinary locations, though none could be interpreted as directly representing the interests of the twins.

Yet many of them expressed concerns about how the twins would think and feel about the research they were about to approve. Zdenek Hrubec, the NRC staff member who was in charge of the registry, balked at the first proposal, saying that "no one could foretell what further applications for access to the Registry might be made in the next few months, and it would be undesirable to wear thin the patience of the twins when it might be necessary to return to them again soon." The twins would need to be informed clearly of the goals of the prospective study (in this case, on the effects of smoking and air pollution on health) and told that they might need to fill out another questionnaire very soon, so that they would not be "unpleasantly surprised."[32]

A second proposal, for a study of twins discordant for schizophrenia, would seem to have been precisely the kind of study for which the registry was intended. Schizophrenia was a well-established "twin" issue, and the protocol for the study was consistent with prevailing research norms and dealt with an issue that seemed to interest many investigators. Yet committee member Paul Lemkau, of the Johns Hopkins University School of Medicine, wondered "whether it was proper to violate the confidentiality of the information available to the Agency by disclosing to a third party the fact that a veteran had been diagnosed as schizophrenic." The VA had, of course, already disclosed the diagnosis of schizophrenia to a third party—the NAS-NRC twin registry—and broad institutional access to information about the private medical experiences of veterans, including psychiatric disease, was a precondition for the entire Follow-Up Agency program. Such knowledge had to move from the clinical encounter to the research database in order for the registry to exist. The solution to the delicate question of privacy proposed at this meeting was that the twins who had been diagnosed with schizophrenia, or their caregivers, should be called by the NRC staff to request permission for their names to be made available to any researcher interested in the disease.[33]

Schizophrenia was clearly "private," in the eyes of committee members. But was residence in the Washington, D.C., area private? One proposal would require information about all twins living in the Washington area. In a "general discussion," according to the minutes, members debated the "propriety" of releasing information without consent and the "problem of possibly exhausting the good will of the twins with the proposed examination, so that other investigators in the Washington area might be debarred from useful access for some years." Confidentiality seemed less pressing in this case—because no stigmatizing disease was involved. But annoyance seemed like an important issue. The "concentration of medical investigators in the Washington area" meant that "many groups might wish to have access to the veterans" there. It did not seem reasonable to preclude the use of the twins in this location for some unspecified time, nor did it seem appropriate to "obstruct useful investigations merely in the hope that someday something else might turn up."[34] In these early meetings, then, participants frequently expressed reservations about research projects that might disturb the twins.

When the "principles which govern the use of the NAS-NRC Registry" were drawn up a year later, four of the five guiding concerns dealt with the responses and vulnerabilities of the twins. All requests for use of the registry, the draft suggested, should be assessed in the light of the need to safeguard twins' privacy, determine that no embarrassment or harm would come to them, protect them from too frequent solicitation, and coordinate the work of investigators interested in the same or related questions. Only one of the guiding principles was explicitly technical: "to insure the scientific value and practicability of the study intended." There was a general consensus that all contacts with the twins should be through the registry itself and that a name and contact information would never be given to a researcher until that twin had specifically agreed to participate. The "budget of good will and willingness to cooperate" of the twin members of the registry was presumably limited, and no study would be approved that might diminish that good will. Any questionnaire or form of physical examination that might alienate the twins should be rejected.[35]

By 1972, twin studies were in progress on smoking and air pollution, schizophrenia, heart disease, dermatology, multiple sclerosis, genetic aspects of headache, and various methodological questions relating to twins, such as the environmental similarity of monozygotic and dizygotic twins.[36] The NAS-NRC committee had by then considered twenty study proposals and approved thirteen.[37] As the twins themselves began to ask questions about the research program, the NRC staff decided that a formal report to the twins was needed. Twins had written in saying that they wanted to know what scientific articles had been published, and they wanted to know why some of the studies were being carried out. They had queries about some of the questionnaires, asking about the purpose of the research and the conclusions reached. In 1971, Hrubec wrote a letter and report for the twins, giving them a very brief explanation of all the research conducted and a list of publications. He considered this notice to be a "compromise between specific reference citations, which would be meaningless to most of the twins, and a listing of subjects reported on, which would not permit those who are interested to locate the reports."[38]

The projected and imagined feelings and responses of twins were thus threaded through the discussions of the use of the registry, and the emotional impact of every proposed research project on the twins was part of the negotiation over its legitimacy and appropriateness. At one level this might seem obvious or even trivial—of course the views of research subjects who cannot be compelled to participate matter!—but my point in emphasizing it is to call attention to the degree to which such expectations by the overseers of the registry shaped decisions about which projects to approve and which to reject. I mean to suggest that if we wish to think clearly about how biomedical knowledge is made, we have to think about these delicate social negotiations, which play a major (increasing) role in biomedical research.[39] We have to think about the fragile consensus in which research subjects often participate and to recognize that many different kinds of knowledge and different social locations are expressed in the highly collaborative text of the scientific paper.

By the 1990s, more than two hundred scientific papers had been published that drew on the Veteran Twin Panel, on topics as diverse as bone mass, eye disease, personality, antisocial behavior, cancer, suicide, Alzheimer disease, headaches, body fat distribution, and alcoholism (Berkowitz and Santangelo 1999, 45). Some of the projects employing the veteran twins moved beyond the social worlds and psychological states of the twins to bring capitalism, Wall Street, and annual compensation packages to bear on questions of nature and nurture. "Earnings," a socioeconomic category of profound political importance, came to be explained through the identical bodies of twins, in a project that pushed the twin method about as far as it could be pushed.

$EN(HRS)W = Y = a + b\,ED + c\,Ab + d\,Back + e\,Pers + f\,Exp$

In the fall of 1972, an economics professor at the University of Pennsylvania, Paul Taubman, asked the subcommittee overseeing research on the twins to approve a project that would use twins to determine the "importance of nature and nurture on earnings, occupational mobility, healthiness and family size." Taubman had no biological training, but the subcommittee was "impressed by the proposal's careful analysis of the problem" and by Taubman's apparent understanding of the twin method.[40] Taubman had already done some work on the question of earnings and innate ability. In an earlier study, which drew on a different group of research subjects mostly under age thirty, he had concluded that "one-third of the gross returns to education" were attributable to innate "ability" in the more educated. Unfortunately this research had been hampered in two ways: his subjects were young (and their long-term earnings therefore unclear) and innate ability was determined by a "small set of tests on which performance may be influenced by differences in the quantity and quality of education and in environment prior to taking the test." In other words, it was difficult to determine, based on testing, how much of the ability was innate and how much was learned. The veteran twins, he proposed, provided much more robust and reliable research subjects. Determining the relationship between innate ability and income in identical twins would be simpler because the precise level of innate ability need not be determined, Taubman said. It was enough to know that they had the same innate ability. Comparisons of their incomes, then, could provide a guide to the correlation between innate ability and tax bracket.

By this logic, if identical twins reared together had the same level of education, their income should, on average, be the same; if their incomes differed, they would presumably differ in educational levels, and they would therefore "provide the closest approximation to a controlled experiment on the monetary returns to education." Average income, ordered by education, in identical twin pairs could provide powerful statistical evidence of the precise quantitative value of education (8).[41] Taubman thought fraternal twins, too, should have similar incomes if they had similar levels of education, but greater variance in earnings in fraternal twins would be still another sign of the importance of heredity.

Taubman knew his model was too simple and that earnings were influenced by more than two factors, but he proposed that it would be "easy to modify the analysis to include other determinants and variables by using multiple regressions." Twins could have been affected by "such things as their wife, number of children, illness, etc." He also knew that twins were good at producing suggestive data, and even though the "so-called heritability indices" could not be treated as "sacrosanct truth," they could still be compelling enough for economic analysis. In biology, such limitations might matter, but in economics, he suggested, they did not (11).

He had an equation to capture his theory:

$$EN(HRS)W = Y = a + b\,ED + c\,Ab + d\,Back + e\,Pers + f\,Exp$$

where EN is the fraction of weeks employed, HRS is the number of hours worked per week, W is the average wage rate, Y is before-tax earnings, ED is the index of quality and quantity of education, Ab is the set of innate abilities that are genetically determined, $Back$ is the set of background factors such as family wealth, *Pers* is the set of personal characteristics such as health, and *Exp* is the length of work experience (15). Taubman's equation plugged heredity right in with hours worked and before-tax earnings. Genes were embedded in the income calculus.

His proposal tacked on an even more surprising claim. He suggested that twin brothers were so thoroughly "the same" that their familial differences could be used to understand and quantify what he called "wife effects." Studies of family size and child rearing, he said, were complicated by the fact that decisions were made "by both husband and wife." Twin brothers, "in terms of tastes and attitude and to a lesser extent earnings and education," could be considered to be "the same." Differences in the number of children and in child-rearing practices could therefore be used to demonstrate the effect of wives on demographic issues. "We plan to examine the number and spacing of children in relation to income, cost of children including earnings foregone by wife, and various education and socioeconomic variables of husband and wife" (19).

Taubman's proposal was approved, and he was granted access to the veteran twins to study the genetics of earnings and wife effects. He may have benefited from his transparent status as an outsider to genetics whose enthusiasms could be forgiven. But he also benefited from his status as an insider at the NRC. Taubman was the chair of an NRC committee on higher education, and the chairman of the NRC's Board on Human Resources made a formal appeal to Hrubec in support of Taubman's proposal. Robert Morse said that Taubman's work was precisely the kind of interdisciplinary work to which the NRC was committed.[42] The Follow-Up Agency did not give Taubman any money—the money was to come from the National Science Foundation and the Hoover Institution—but it did approve his access to the veteran twins and did work with him to develop a questionnaire and mailing program.

In 1974, Taubman sent a survey to a thousand monozygotic twins and a thousand dizygotic twins in the NAS-NRC twin registry. These survey results became the evidentiary basis for Taubman's 1977 paper on the "determinants of socioeconomic success," in which he wrote that "the extra earnings derived from education are much smaller than in other studies which have not controlled for genetic endowments and family environment." Taubman's "earnings phenotype" was heavily shaped by heredity: 45 percent of earnings were attributable to genes, 12 percent to family environment, and the remainder to other factors external to the abilities of the worker. "Thus even with complete equality of opportunity," he concluded, "inequality of outcomes would be reduced by only 12 percent" (Taubman 1978, 177; see also 1976a, 1976b). It was a classic statement of the hereditarian perspective, embedded in a paper filled with equations: improving educational opportunities and the human environment would have cash value neither for the worker nor for society. As Alan Gregg put it in 1945, "wise matings" were better than "$800,000 high schools" (quoted in Paul 1991, 262).

The surreal qualities of Taubman's use of the twins coexist with a familiar, down-to-earth, practical emphasis on social efficiency and cost-benefit calculation. Taubman replicates a fundamental structure of twentieth-century technocratic rationality, as clearly expressed in his work. He wanted to know whether education increased earnings. From his perspective, the difficulty with studies that did not use twins was that the well-educated and well-paid worker might have started out with a biological advantage, which might call into question any conclusions about the cash value of education. Identical twins held heredity constant—they made it possible to assess the (minimal!) impact of education on income, or even to measure the impact of the "wife effect," because the twins were essentially copies of each other. They were measuring devices, instruments through which the power of both heredity and environment could be calibrated.

There were other twin studies that dreamed of simplicity in the 1960s and 1970s. One paper tracked marriages of twins to twins and found 100 percent concordance for either happiness or divorce, concluding that "it may become feasible to develop pre-marital screening tests based upon chemical and physiological measurements, to predict certain cases of gross incompatibility" (Taylor 1970, 97). More notoriously, Arthur Jensen used twins to demonstrate the strong heritability of IQ, a commodity of transcendent value in the frames of technocratic rationality. Jensen was under the impression that he had found a biological explanation for racial inequality, an impression he shared with the world in 1969 (Jensen 1969, 1974).

But I want now to leave simplicity behind and to turn in my final discussion to studies that dreamed of complexity and that drew on veteran twins for genetic explanations of precisely the kind of complex behavior most vexing in public health and clinical medicine and most likely to demonstrate the practical value of human genetics to the practicing clinician.

The Gene for Alcoholism

Alcoholism is a behavior the average family physician presumably encounters far more regularly than schizophrenia or genetically driven low income. A meat-and-potatoes disaster, alcoholism is spread throughout income levels and races, and it poses serious problems of both physical and mental health and public policy. An explanation for alcoholism—especially an explanation that could facilitate its control—would be dramatic evidence of the value of the new human behavior genetics to public health and clinical medicine.

Alcoholism has been a tantalizing genetic end point. It runs in families, seems to have some biological underpinnings, and is as stubborn and difficult to change as any "fundamental" quality of the body. Without exception, family studies of alcoholism in many different cultures have shown higher rates among the relatives of alcoholics than in the general population, for as long as such studies have been done (Goodwin 1979). Complete abstention from alcohol use runs in the same families, apparently as a social response to the experience of interacting with an alcoholic. Because simple pedigrees of alcoholism made a powerful case for heredity, alcoholism was a candidate for genetic explanation for the eugenicists, the racial hygienists in Nazi Germany, twins researchers in the 1950s–70s, and molecular geneticists in the technologically sophisticated final decades of the twentieth century. Yet a genetic explanation remains elusive.

Twins have been used in studies of the genetics of alcoholism, but these studies have been complicated by the use of *alcoholism* to cover a wide range of behaviors and by persistent questions about zygosity in large samples of twins. For example, when the twins in the NAS-NRC Veteran Twin Registry were used in a study of alcoholism, the diagnosis came only via the computerized records of the VA hospitals, and the very large sample—15,492 white male twins then between the ages of fifty-one and sixty-one—had been assessed for zygosity almost entirely by questionnaire. The combination of uncertain diagnosis of both disease and zygosity led one later analyst to suggest that the report was "so lacking in firsthand information regarding the

---

Taylor's twins married to twins (Taylor 1970).

drinking habits of the majority of the sample that it is impossible to make any inferences from it about the genetics of alcoholism" (Marshall and Murray 1989, 280).

Some earlier twin studies had involved more direct data gathering. The Swedish psychiatrist Lennart Kaij studied 174 male twin pairs who were identified through county temperance boards, which received reports of alcohol abuse throughout southern Sweden (Kaij 1960). He interviewed 292 of the 348 twins. That is, he spoke to them, went to their homes, and took a social and drinking history. He determined zygosity by looking at the twins and measuring them. Sometimes he got the two twins together and compared them side by side. In a few cases, he did blood typing (Marshall and Murray 1989). Like many others, Kaij found higher concordance rates for alcoholism in monozygotic twins, though how much higher depended on his definition of alcoholism.

And could the twins, could anyone, be trusted to report the true nature of their drinking problems, or any of their problems, even in the setting of a VA hospital or a private interview with a psychiatrist? In 1976 Loehlin and Nichols tabulated results from twin studies showing that relying on self-reports for behavior invariably found monozygotic twin correlations approximately .20 greater than dizygotic twin correlations, no matter what behavior was being assessed (Loehlin and Nichols 1976, cited in Marshall and Murray 1989). Twins were being asked to describe their fetal relationships, their inner mental states and true selves, and their uses of a substance with highly charged social meanings, alcohol. Could they be made to speak for nature, even in the aggregate?

In 1990, a gene for alcoholism was reported with much fanfare in the *Journal of the American Medical Association* (Blum et al. 1990). Frozen brain cortex tissue samples from thirty-five autopsied alcoholics and thirty-five nonalcoholics were compared at the locus of the human dopamine receptor gene on chromosome 11. The alcoholics were found to be more likely to have a particular allele, A1. This finding was not replicated, other alcoholism-related chromosomes were identified (1 and 7), other genes and biological mechanisms proposed, and multiple typologies of alcoholism (types I, II, and III) brought in to make sense of the scattered data. The genetically puzzling fact that for every female alcoholic there are three to five male alcoholics led to proposals that male alcoholism was hereditary and female alcoholism was not,[43] but the genetic mechanism was unclear. A 2000

Handbook of Alcoholism intended for health care providers was diplomatic: "Family and twin studies provide strong evidence to support genetic transmission of alcoholism, but also indicate an equally important role for shared environment" (Zernig et al. 2000, 305).

My point here is that alcoholism is a syndrome, or cluster of behaviors, or psychosocial strategy, that has aroused intense desire in the community of researchers working on twins, families, and behavior genetics. Dozens of books, papers, symposia, and conferences have been devoted to the genetics of alcoholism over the past forty years (Kaij 1960; Partanen, Bruun, and Markkanen 1966; Seixas et al. 1972; Goedde and Agarwal 1987; and many others). Research seeking the gene for alcoholism, including major initiatives at the National Institute of Mental Health, has continued despite repeated setbacks. The search has continued despite the recognition within the behavior genetics community that alcoholism is heterogeneous, despite the retraction of various found genes, and despite the lack of a compelling biological model for how this particular end point—alcoholism—is produced.

The discovery of a gene for alcoholism might not benefit alcoholics—Stanton Peele and other critics have elaborated the possible psychological and social consequences of the applications to which a gene might be put—but it would certainly benefit molecular genetics, which could then lay claim to the ultimate quarry, the gene that explains a complex, environmentally mediated disease important to almost everyone.[44] And with a gene identified, alcoholism, once embedded only in the corporate nexus of the liquor industry, could be repositioned in the world of biopharmaceuticals and prenatal testing, where it could become the site of even more profitability.

Conclusion

On its website, the Institute of Medicine of the National Academy of Science states that the Veteran Twin Registry is at a point of maximum usefulness as the twenty-first century begins, because of the "rapid accumulation of morbidity and mortality endpoints in this population of men aged 69 to 79" (www.iom.edu). An estimated five hundred men in the Veteran Twin Registry die annually, however, and half of the pairs are no longer intact. The NAS is therefore beginning an effort to develop a Current Era Twin Registry enrolling military twins "indefinitely into the future." To be born a twin is to be scientifically useful in perpetuity, and the high technology of twenty-first century molecular genetics does not eliminate the need for bodies emerging from the same womb.

Indeed, the creation of twin registries escalated after the 1960s, with the support of the National Institutes of Health, universities, state governments, and private foundations. The Virginia Twin Registry was developed at the Medical College of Virginia by Linda A. Corey and Walter E. Nance in 1978—with the help of a "one page zygosity determination questionnaire." It merged in 1998 with the North Carolina Twin Registry to create one of the world's largest twin registries, with thirty-seven thousand pairs, the Mid-Atlantic Twin Registry (www.matr.gen.vcu.edu). This registry has received funding from the W. M. Keck, Robert Wood Johnson, McArthur, and John Templeton foundations (more than $4 million total) and the National Institutes of Health. Many institutions have a stake in twins research. Meanwhile the National Institute of Mental Health supported the creation of a twin registry at the University of Wisconsin (seven hundred pairs). The Northern California Twin Registry, created in 2000 by the metamorphosed Stanford Research Institute (now SRI International), relies on advertising and recruitment, rather than birth records, and includes, as of January 2005, approximately two thousand registrants (www.sri.com/policy/healthsci.twin). The largest solely recruited twin registry is the Australian Twin Registry, with more than thirty thousand pairs by January 2005; it was created in 1981 and enrolls twins of all ages and states of health (www.twins.org.au). This registry has generated more than two hundred peer-reviewed scientific papers (Clifford and Hopper 1986). Collecting twins is generally an industrialized activity, but there is one twin registry in the developing world: the National Twin Registry of Sri Lanka, created in 1997 (www.infolanka.com/org/twin-registry). And some of the newest registries focus only on twins with particular disorders—for example, the University of Washington is developing a registry of twins one or both of whom have chronic fatigue syndrome or fibromyalgia (http://depts.washington.edu/uwccer/uwtwinsr.html). There are also twin registries in Belgium, Canada, China, Denmark, Germany, Finland, Japan, Korea, and Norway. Many of these collections of twins are described on the website of the International Society for Twin Studies (www.ists.qimr.edu.au), and most also have their own web page, linked to the society's page and used to recruit both twins and scientists. The twin is the original clone, "worth millions" when compiled in costly registries built around one-page questionnaires. A quick scan with an internet

MOMENTS OF TRUTH IN GENETIC MEDICINE

search engine turns up many sites for such twin registries. All use the questionnaire method for the diagnosis of zygosity.

Sandra Scarr, in her presidential address to the Behavior Genetics Association in 1986, presented a triumphal narrative of knowledge overcoming ignorance in the IQ debate (Scarr 1987).[45] She then appealed to an unusual source of scientific legitimacy. Regarding the intense environmentalism underlying John Broadus Watson's behaviorism, Scarr said that "somehow parents of more than one child knew this had to be wrong. (I am often amused that the most ardent advocates of the dramatic effects of parenting are either nonparents or the parents of only one child.) Parents of two or more children know perfectly well that their children are different for reasons that have nothing to do with their training regimens. Most parents know that their children evoke different regimens from them and that they have limited effects on their children's development." She added that "there has always been a reservoir of sympathy out there in the public consciousness for what we have been promoting" (227). The common folk knowledge of human nature—the knowledge of parents of two or more children—was thus invoked to suggest the legitimacy of the enterprise of behavior genetics. The families and friends of twins could discern zygosity, and the parents of two or more children could discern the hereditary forces underlying their children's behavior. Behavior geneticists just came in to do the numbers on what was already known.

I am interested in this seamless transfer of knowledge from the "parents of two children" to the scientific text partly because Scarr is not alone. Many essays and behavior genetics textbooks open with a discussion of the commonplace, everyday sense that people are born with some aspects of their personalities and abilities in place, and many of the websites for twin registries express the same idea. These widely shared feelings and observations have been an important resource for the field of behavior genetics, as participants have drawn on their own experiences—their own sense of a fundamental, changeless self—to explain their discipline. These narratives express what I would identify as a modern, post-Enlightenment sense of the inflexible self who cannot react quickly enough to fluid social environments: people's inborn changelessness becomes a problem to be explained only in social worlds that require flexibility and adaptability.

Twins' perceptions were integral to the technical descriptions of genes for schizophrenia, alcoholism, earnings potential, and manic depression, to the structure of the research approved for the NAS-NRC Veteran Twin Panel, and to the assessment of a covert fetal event, zygosity. As the twins were brought together in registries, they participated actively in the behavior project, filling out multipart forms, taking psychological tests, enduring lengthy physical exams, and providing their life stories (employment history, marital situation, experiences in war) to the makers of technical knowledge.

The accounts that twins gave of themselves to the psychiatrists, geneticists, and others who studied them conformed to the standardized, historically specific frames of psychological testing. An interior world of meaning and logic was extracted through such tests, through the lists of traits twins were asked to claim or reject for themselves ("infantile" or "good-looking" or "vindictive").[46] Twins were told to mark the adjectives that described them. "Try to be frank and fill the circles for the adjectives which describe you as you really are, not as you would like to be." The twins could tell the scientists how they really were, not how they would like to be, and this extracted interior world, this mental vision of the true self, could be used to reach conclusions about heredity. Twins told researchers whether they were identical, and they told researchers what genes they shared. They spoke the narratives of heredity to the technical experts.

Knowledge of the genetics of human behavior thus often included folk knowledge, and scientific papers in behavior genetics were often an expression of the interior mental worlds of many people, including twins and those who knew them. Genes were products, in a very direct sense, of the twins autobiographical constructions, hypothesized based on the twins' mental worlds, as those worlds were expressed and made manifest in intelligence tests, personality tests, accounts of social reactions, reports of childhood experiences, sensory response tests, and measurements of neuroticism and extraversion.

The disturbing possibility that culture might be inside the brain—that it might be expressed by the twin, brought out and placed on the page in the standardized tests, and played out in interviews about drinking behavior or shyness or aggression—was threaded awkwardly through this scientific literature, almost from the earliest papers on twins. Did twins know their true inner selves? Did their words explain alcoholism? Did they tell the truth? And did they really know their own zygosity? For behavior geneticists, twins provided one of the most compelling and reliable entrees to genes by seeming to black-box heredity by their circumstances of birth. But twins were reg-

MOMENTS OF TRUTH IN GENETIC MEDICINE

isters of both culture and nature, and when they described their lives and their fetal origins, they were inside the culture of the inflexible, solid, inborn self, the core identity that could not be changed, the "human nature" that Scarr proposed every parent of more than one child recognized. The twins were collaborative participants in the massive, multi-institutional project to find that changeless self in heredity.

CHAPTER 6

[JEWISH GENES]

History, Emotion, and Familial Dysautonomia

Most genetic diseases are "small." They affect very few people. They are also often strange and difficult to recognize. Most physicians never see most of the diseases that appear in Victor McKusick's catalogue, *Mendelian Inheritance in Man*. Unlike cancer or heart disease or depression, these diseases have a limited effect on everyday medical care. They can be devastating for the individual and family affected, but they are not important, or familiar, or perhaps even worth mapping to the genome.

In this chapter I look at one of those diseases, familial dysautonomia, that most physicians will never see and that most genomic scientists initially had no interest in mapping. Indeed, although it might have its origins in a single fifteenth-century mutation, it is a disease that was not clinically visible until the twentieth century. The new antibiotics of the 1940s permitted the very vulnerable infants born with familial dysautonomia (FD) to survive long enough to be noticed and to be differentiated from the mass confusion of infant mortality: FD is a postantibiotic disease, experienced and lived as a result of a new therapeutic technology.[1] It is also a disease that may well disappear, also through technological intervention. A genetic test was first developed in 1993, and in early 2001 the gene was found. Prospective parents who are at risk (and the relevant risk factor is Ashkenazi Jewish heritage) can be tested for carrier status, and fetuses can be tested prenatally and aborted if they have the disease. Many parents, physicians, genomic scientists, and others want to prevent the birth of more babies with FD. It could, perhaps, become a disease for which the generation of affected people alive today is the last. It could become a disease only of the twentieth century, efflorescing from 1940 to the end of the century, then disappearing. And this disappearance, too, will be the result of technology.

I want to hold in sight throughout my discussion the notion of a disease both brought into being and (to be) eliminated by social organization and technology. I want the quality of FD as constructed to be central to my narrative, which draws primarily on the scientific texts that defined the disease and explained its meaning. Familial dysautonomia, as an experienced and lived disease, is a product of neuronal anomaly, of sensory dysfunction, of Jewish history, and of biomedical technology. It is a "natural" phenomenon, a splicing mutation, named DYS, in the IKBKAP gene, a mutation that causes a tissue-specific absence of exon 20 in the IKAP protein within the brain. And FD has been sustained and brought into existence by Ashkenazi Jewish reproductive practices. It is also an extremely demanding bodily state for those affected and their families. FD is an amalgam of history, emotion, biochemistry, and medical technology.

Familial dysautonomia was noticed because the Ashkenazi population in New York City was dense enough to produce many cases in a relatively short period. In the 1940s, three physicians at Babies Hospital in New York encountered five "puzzling cases." All these children did not produce tears. They suffered from intense vomiting and experienced skin blotching, sweating, blood pressure instability, and other signs of "disturbances in autonomic function." In an atmosphere of "continual exchange" within the hospital's social network, physicians Conrad Riley, Richard Day, and their colleagues "recognized the similarity between the patients," and in the spring of 1948 they prepared a report suggesting that all five children had the same condition. The syndrome, described as "bizarre" and "hitherto unrecognized," came to be known temporarily as Riley-Day syndrome (Riley 1970; Riley et al. 1949; Brunt and McKusick 1970).

That the disorder might be hereditary was considered possible from the beginning. It was present from birth, and all those affected were Ashkenazi Jews and therefore members of a genetically distinct population. In 1954,

Riley concluded that the disease was genetic and renamed it familial dysautonomia (Riley, Freedman, and Langford 1954), but it continued to be called Riley-Day syndrome in scientific publications well into the 1970s. Over the next twenty years, more than two hundred cases were reported in the medical literature. By 1964 a reliable diagnostic skin test (a characteristic muted response to histamine) was available. A special clinic for the study and treatment of familial dysautonomia was established at New York University in 1970, under the guidance of a young physician, Felicia Axelrod, who has since devoted her medical career to the treatment and management of FD (New York State Department of Health 1980; Axelrod 1997a, 1997b).

But what exactly was the disease? Its experiential and clinical manifestations were baffling, and affected children had many complicated medical problems. There were obvious dysfunctions in the nervous system, and in the decade from 1969 to 1979 electron microscopy studies by Aguayo and Pearson revealed a decrease in the number of very small peripheral neurons in patients with FD (Aguayo, Nair, and Bray 1971; Pearson, Budzilovich, and Finegold 1971; Pearson, Brandeis, and Cuello 1982). This led to an understanding of FD as a disorder in which the nervous system had not completed its development (Axelrod 1997b, 529). In 1992 the FD parent support group began to fund gene-mapping research by a Harvard University team headed by James Gusella, who was then mapping the Huntington gene. And in the summer of 1993 his research team located the general vicinity of the FD gene and developed a genetic test that could be used for prenatal screening of future pregnancies in families with at least one affected child. In early 2001 the DYS gene was found by two different labs, and a priority dispute broke out in the press, over the internet, and among the parents. Some sided with one lab, some with the other. A splinter parent support group was formed, and long-simmering rifts among the various interested parties erupted in open disagreement.

The public contention, however, coexisted with fundamental consensus. All sides sought to prevent the birth of more children with FD. In Ashkenazi families, because of science, technology, and social organization, another FD child need never be born. Thus, between 1949 and 2001, a previously unknown condition appeared, was named, and was characterized in terms of its "bizarre" bodily and psychic consequences and, eventually, in terms of its precise location in the peripheral neurons and on the distal long arm of chromosome 9 (q31–33) (Blumenfeld et al. 1993; Slaugenhaupt et al. 2001; www.familialdysautonomia.org).

I explore here the history of familial dysautonomia, suggesting that this disease, so anomalous in some respects, can provide insight into more general changes in the status and meaning of genetic disease. I propose that FD is a useful case study for several reasons. First, it is a classic genetic disease, the vast majority of which are rare and unlikely to be seen by most clinicians. Second, it was one of the earliest genetic diseases to acquire a parent support group. The Dysautonomia Association (now Foundation) was created in 1951. Most genetic diseases acquired parent support groups after 1980, though a few prominent genetic diseases had such groups earlier—for example, hemophilia (1948), muscular dystrophy (1950), and cystic fibrosis (1957).

Familial dysautonomia is also a useful case study because of its relationship to a much more famous genetic disease of Ashkenazi Jews, Tay-Sachs disease, which was also transformed in this period, though in very different ways. Both Tay-Sachs and FD became the focus of interest in American medical schools in the 1970s, and both are the result of recessive alleles that are present in Ashkenazi populations at similar rates. One in thirty individuals of Eastern European Jewish extraction carries the recessive FD allele; one in twenty-five individuals of Eastern European Jewish extraction carries the recessive Tay-Sachs allele. Both diseases also have advocacy groups, volunteer support networks with web pages, public information programs, parent resource guides, and fund-raising activities. Tay-Sachs disease is, of course, a well-known, widely publicized genetic disease, and its control through carrier testing is a success story that has been the subject of many media accounts and much medical interest (National Tay-Sachs and Allied Diseases Association 1994; Kaback 1977).[2] Conversely, FD was socially, medically, and politically invisible until early 2001, when publicity about the gene led to extensive press coverage.

Both FD and Tay-Sachs disease are disorders that do not, in their direct consequences, much engage public health debates. The vast majority of genetic diseases individually affect very few people, whatever claims one might want to make about their overall public health impact. Yet these "small" diseases have played a critical role in the broadening and reconceptualization of genetic disease in general.

Seeing the Disease

Familial dysautonomia was difficult to see and to identify as a single problem. In one ten-year-period, between 1938 and 1948, physicians at Babies Hospital and at the New York Hospital in New York saw five children with "symptoms so puzzling as to defy exact diagnosis, yet so similar as to constitute a clinical entity." Conrad Riley and three other New York pediatric specialists wrote a "report of five cases." This report was published in 1949 in the journal *Pediatrics*.

In their paper, the authors tried to extract the common features from their case histories but could offer only two: first, "undue reaction to mild anxiety characterized by excessive sweating and salivation, red blotching of the skin and transient but marked arterial hypertension," and second, "constantly diminished production of tears." Yet the case histories themselves described many other symptoms, and presentation at the hospital was never for the child's undue reaction to anxiety or for diminished production of tears. These were not the manifestations that provoked parents to bring their children to medical attention. Instead, they came because the children were vomiting or because they had corneal ulcers.

These cases thus had little in common, and extracting a syndrome from them was a work of great imagination and insight. Case 1 in the 1949 report, for example, was a girl first seen at age six who had attacks of vomiting, cold hands, a tendency to drool, excessive nighttime perspiration, and retarded mental development. When the paper was published, this patient was sixteen years old and attending the eighth grade. A list of ineffectual drugs that had been administered to her "without appreciable change in symptomatology" included atropine, phenobarbital, ephedrine, nicotinic acid, and benzedrine. Case 2 also came to the attention of physicians at Babies Hospital relatively late. The boy, at the age of five and half, was brought to the hospital because of a vomiting spell that had continued for more than twenty-four hours. He had been hospitalized elsewhere six times since his birth, three times for pneumonia and three times for other conditions including cyclical vomiting. Over the next six years he was placed in a psychiatric institution, treated with phenobarbital and atropine, and subjected to extensive medical tests. In 1945, when he was eleven and staying in a mental hospital, he began to vomit blood, fell into a coma, and died. In case 3, the boy exhibited seizures as a newborn, blood pressure problems at eight months, and developmental delays at one year. His extremely variable blood pressure was documented at Babies Hospital when he was four years old, and he was then diagnosed with "autonomic imbalance." At one point his foot was immersed in ice water, and "surprisingly he complained of no pain" and his blood pressure showed no change. "On questioning, it was discovered that he never cried with tears although he never had any trouble with his eyes." The boy in case 4, unlike the others, was actually subjected to a neurosurgical therapy, rather than drugs, in an effort to alleviate his problems. A "boy of Jewish parentage" whose father had been rejected by the draft board (a detail noted in the published paper), he first came to medical attention at the age of eighteen months because of a corneal ulcer of the left eye. Eye surgery resulted in complications that led to his complete blindness by the age of four. The girl in case 5 suffered from vomiting, obesity, cyanosis as a neonate, and intense emotional reactions to mild stimulation. She died at age ten, and her autopsy revealed an "essentially normal" brain, "both grossly and microscopically" (Riley et al. 1949).

What were the common threads linking these patients? The only symptom clearly present in each child was vomiting. These cases differed in presentation, in the patient's age and mental status, and in the precise nature of the crises that brought the patient to the hospital. There was a tentative and suggestive diagnostic link to the autonomic nervous system, particularly given the odd blood pressure readings, but the patients did not seem, in the text of the 1949 paper, to be dramatically "the same." They all had "undue reaction to mild anxiety" and diminished production of tears, but all had many other symptoms as well. Physicians working with them recognized a fundamental similarity, however, and named a disease. And the mere existence of the diagnostic category made it possible for other children to be understood as suffering from the same thing.

Cases began to accumulate at Babies Hospital, as other physicians recommended that patients who had no tears, vomited cyclically, or seemed to have blood pressure anomalies be sent to Riley and his colleagues. By 1952 "we could generalize about 33 patients" and "the familial nature of the disorder was emerging" (Riley 1970).

On Heredity

Scientific and medical observations about familial dysautonomia in the 1950s and 1960s focused on the systemic effects of the disease, which played out in many bodily sites. A wide range of medical and scientific experts therefore studied these effects and proposed possible means of treatment and amelioration. Because the lack of tears often led to corneal ulcers, ophthalmologists studied the condition and compared therapies (Liebman 1957). Anesthesiologists became involved in the study of FD because complications during anesthesia were common in these patients (Kritchman, Schwartz, and Papper 1959). Experts in psychological development assessed intelligence, experts on senses and the nervous system assessed taste, smell, and autonomic malfunction, and heart specialists looked at coronary malfunction (Forster and Tyndel 1956; Moloshek and Moseley 1956; Henkin and Kopin 1964; Smith and Dancis 1964a). So many things were expressed in the FD body that it was broadly relevant to many fields.

In the 1960s, a cardiologist turned human geneticist, Victor McKusick, and his colleague F. W. Brunt, both at Johns Hopkins University, undertook the first specifically genetic survey of FD. They conducted a systematic survey of families in which words rather than blood samples were collected, noting that "cumulated case reports" had sometimes been used to establish the genetic basis of diseases, but "this method is clearly subject to considerable error." Instead, they pursued all known cases of FD and conducted detailed interviews and examinations when possible. With the help of the parent support group, McKusick and Brunt were able to track down 172 families with 210 affected children, living and dead. The ethnic distribution of these families was "striking." Of the 344 parents interviewed, only one was not of known Ashkenazi extraction, and most traced their ancestry to the same area of Eastern Europe (southern Poland, western Ukraine, northeastern Romania) (Brunt and McKusick 1970, 345). Brunt and McKusick wondered whether physicians might be reluctant to diagnose FD in any child who was not of Ashkenazic extraction, but noted that the "clinical picture is so distinctive" that few cases would be missed on this basis alone (348).

The patterns of inheritance in these familial accounts suggested that FD was an autosomal recessive. No parents were affected. There was a relatively high rate of cousin-marriage (5.3 percent). There were also reports in the literature of minor abnormalities in the presumed heterozygotic parents (Moses et al. 1967; Riley and Moore 1966). In their effort to explain the persistence of the FD gene in Ashkenazi populations, Brunt and McKusick explored the possibility of a heterozygote advantage. Heterozygote advantage was a popular hypothesis at the time for deleterious genes that persisted despite their adverse effects on reproduction. FD would have reduced fertility in affected families, simply because most of those born with the disease were unlikely to survive. Yet the gene for FD had apparently been preserved and maintained in the Jewish population. The interpretation of sickle cell anemia in the 1950s suggested that, for some serious genetic diseases, there were advantages for the carrier (said to have sickle cell trait), if not for the affected person. It became clear to biological anthropologists and geneticists that persons with sickle cell trait were resistant to malaria. The disease itself was disabling and serious, but carrier status provided a biological advantage in areas where malaria was endemic. This explained, in a compelling way, the persistence of a gene that caused serious illness, and geneticists and anthropologists began to seek other examples of this phenomenon.

For FD, there was strong geographical evidence, based on their interviews, for a founder effect, a single mutation during or before the fifteenth century in a group of Jews who later migrated to Poland and Russia. But once the gene existed, its rapid spread and persistence, despite its apparently adverse effects on reproduction, demanded an explanation. Brunt and McKusick proposed that fertility might be higher in heterozygotes, though their numbers were too small to be statistically significant (Brunt and McKusick 1970, 350–51), and like other commentators before them, they wanted to "ascribe all the phenomena in this disease to a unitary biochemical defect, a single enzyme deficiency." But the disease, while clearly a distinct clinical entity, had not yet been linked to any specific biochemical abnormality and its etiology remained unclear. Perhaps, the authors said, patients with FD could help "in building up our knowledge of the human nervous system" (369).

The patients (presumably their bodies rather than their minds) would help scientists gain access to more general knowledge of the human nervous system, both by their words (the narrative evidence that McKusick used) and by their biochemical, developmental, or neuronal differences. The patients' bodies and families were a resource for the knowledge-makers. But could they also make knowledge of their own? In their closing acknowledgments Brunt and McKusick thanked those who had been directly involved in the research (medical students who conducted interviews, colleagues who provided advice and comments) as well as "the patients and their parents for their tolerant participation in tedious studies" and the Dysautonomia Foundation, which funded part of the research—the parent support group.

The Parent Support Group

The Dysautonomia Association, Inc., organized in 1951, was one of the first parent support groups for a genetic disease in the United States. The frequency of a disease does not seem to play an important role in the establishment of such groups. FD affects very few people and had an organized group by 1951; Down syndrome is one of the most common congenital disorders in the United States, and it did not have its first parent support group until 1966 (it now has four). I assume that medical attitudes about the condition affect whether parents want to meet other parents, though I am not sure how this plays out in all these cases. In recent years parent support groups have developed for extremely rare conditions (moyamoya disease, Moon-Biedl syndrome), contested conditions with ambiguous medical and social status (fragile X syndrome), and conditions that attracted the attention of genomics researchers before they attracted much clinical attention, because of their location on the genome (velo-cardio-facial syndrome). All these groups are now part of a larger Genetic Alliance (or, before 2000, Alliance of Genetic Support Groups), created in 1985 after a March of Dimes Birth Defects Conference at which the idea of an omnibus organization was widely supported. The alliance has published detailed guidelines on informed consent, a guide to starting a genetic support group, various directories, and many reports on pending legislation and issues relating to patient care (Alliance of Genetic Support Groups 1995; J. O. Weiss and Mackta 1996).

Family support groups have played a critical role in the production of knowledge about genetic disease since 1945. The bodily knowledge of parents, the knowledge acquired by intimate contact with a child whose body does not conform to expected patterns, was incorporated into scientific and medical interpretations of FD. Sometimes the role of parents was explicitly invoked in scientific texts, perhaps in the form of expressions of gratitude to affected families. But often it was more oblique. Long passages in the passive voice in some scientific reports were clearly derived from parental observations of everyday life with a child with FD. "Unusual spontaneous limb movements were seen both at rest and during periods of inactivity," one paper in *Pediatrics* noted in the midst of a detailed description of an infant's behavior before the first clinical examination. Who saw these movements? And who experienced the "enervating quality" of the nasal cry of the newborn (Perlman, Benady, and Saggi 1979)? In another paper, "many parents

MOMENTS OF TRUTH IN GENETIC MEDICINE

volunteered the information that the eyes were not absolutely dry" (when their children cried) (Liebman 1957, 190). Parental descriptions of the FD body provided clues to the underlying phenomenon and could be folded into scientific descriptions of the condition. Their testimony was part of the technical infrastructure.

In addition, parents and families affected by FD contributed funding to most of the research conducted on this disorder, and they contributed bodies, blood, and words (family stories, personal stories) to all such research. They were both an organized physical and social resource and an independent funding agency promoting interest in the disease. When the decision was made to seek the gene, the parent group set the scientific agenda and found someone willing to pursue it, in a convergence of technical and emotional knowledge.

The strength of the support group for FD was shaped by the fact that it is an ethnically limited disease. All children known to have FD have been Ashkenazi Jews or had Ashkenazi ancestry. Indeed, FD's status as an Ashkenazi disease is a defining characteristic. A child with no Ashkenazi connections who presents with a sensory disorder will not fall under FD proper by medical agreement. At least seven other debilitating genetic diseases occur primarily in Ashkenazi Jews: Bloom syndrome, Canavan disease, Gaucher disease, mucolipidosis IV, Niemann-Pick disease, Tay-Sachs disease, and torsion dystonia (National Foundation for Jewish Genetic Diseases 1980; Goodman 1979). Such diseases have an important social dimension as specific to a relatively well-educated group with significant intellectual and economic resources (on other socially specific genetic diseases, see Clark and Hughes 1992; Vullo and Modell 1990).

Conrad Riley, the physician who first described FD, actually created the parent support group, with the help of social worker Marjorie Eustis Frankel and psychiatrist Alfred Freeman. It was Riley's idea to bring parents together so that they could help him understand the disorder, which he tended to witness only when the children were in crisis. Parents might recognize patterns in their children's behavior outside the clinical setting that could shed light on the underlying problem. And so parents became his informants, and Riley credited this parent group with permitting him to "put together a fairly complete description [of FD] in 1957."[3]

Interacting with a profoundly different sort of body—the body of a child with FD—as a parent would interact would provide, in the process of caring labor, knowledge not present or accessible in either clinic or laboratory. Riley needed to know the intimate, day-to-day patterns of the disease in order to understand it. And parents, particularly mothers, had that kind of knowledge, though they might not interpret what they knew as medical knowledge.

The explicit function of the parent support group as a network for the production of knowledge was articulated in a 1956 pamphlet that the group published "as a public service," entitled *Living with a Child with Familial Dysautonomia* and written by Riley. "The purpose of the booklet is to give fathers and mothers of these handicapped children some understanding of the kind of problems to be faced . . . it is the outgrowth of the work of a group of doctors, social workers, psychologists, speech therapists and, above all, parents of such patients. From this 'team' has been assembled the information here included" (preface). The parent support group, Riley wrote, was begun "for the purpose of helping parents to understand better the nature of their children's disorder," and this had been accomplished, but "in addition the meetings have proved a great source for increasing the doctors' understanding of the condition" (37). The final text, Riley suggested, was a collaboration, a distillation of these meetings, with their give and take between doctors and parents, social workers and psychologists.

Riley first described the autonomic nervous system and the symptoms and complications associated with the disease—in other words, the matters relevant to clinical medicine. But in chapter 6 Riley turned to "general management," and here his informants were parents. He proposed that the adults caring for a child with FD should seek to protect the child from "situations requiring adaptation," because children with this disorder found it difficult to adapt—to react "in a controlled fashion to outside stimuli." Parents should avoid immunizations, which required and provoked a bodily reaction. So, too, they should seek to limit the child's exposure to frustration, because frustration provoked a psychic reaction that became, in children with FD, a dramatic physical reaction. Yet the complete avoidance of "any frustrating situation" would be not only impossible but counterproductive. The normal tension, he said, between parent and child was exacerbated by the intense emotional responses of children with FD: "Failure of the parent to accede to baby's wishes may cause a major emotional upheaval, perhaps to the extreme of breath holding until he loses consciousness" (28). Terrified by such incidents, parents sometimes chose to make concessions to the child at "every point of difference." But this only delayed the crisis, which could then take "the form of a severe vomiting attack which requires hospitalization." The standard pattern of children resisting parental control could become, for the child with FD, a major medical event. Instead of leading to a tantrum or withdrawal of privileges, it could lead to hospitalization.

The qualities of the disease itself bound together technical and emotional management. The incomplete development of neurons, the absence of a molecular growth factor at a critical point in the maturation of the fetal nerve cell, produced a child whose emotional reactions to parental correction could lead to the emergency room. The body-mind continuum was manifest dramatically in the FD body, where thoughts became vomit and the bodily knowledge of the caring parent became scientific knowledge in a formalized text.

The Psychic State of the Patient with Familial Dysautonomia

The question of how this mind-body continuum could be situated within the disease was the subject of a series of papers after 1956. Language difficulties, intelligence, and emotional maturity were assessed partly as clues to the underlying physiology of the disease and partly as guides to clinical management (Lawrence 1956; Freeman, Heine, and Havel 1957; Sak, Smith, and Dancis 1967; Garwood and Augenbraun 1968).

In the 1970s, Axelrod, David Clayson, and Wooster Welton tried to understand the psychological dimensions of FD, and their two papers on the subject in *Pediatrics* normalized patients with FD by arguing that they were neither retarded nor psychologically incapable of attendance in regular school programs. The scientific agenda of disease creation, in this case, was turned to diminishing the otherness of the "bizarre" disease, making it something manageable rather than strange. They found "less cognitive impairment than previous research would suggest"—in other words, patients with FD were of normal intelligence and could adjust to standard schooling. Children with FD did seem to have some intellectual deficits, including lower scores on tasks that required them to manipulate objects. Yet "the same proportion of the dysautonomic population scored within the average range of intelligence as is found in the general population." There were certainly more lower scores in the FD population. This the researchers attributed to the difficult emotional management of the disease, which could interfere with learning and attendance (Welton, Clayson, and Axelrod et al. 1979; Clayson, Welton, and Axelrod 1980).

Earlier studies of intellectual and psychological development in children with FD had been contradictory. A 1956 study of a single twelve-year-old patient suggested that intellectual development was normal (Lawrence 1956), and this was backed up by a later, larger study of fourteen patients with FD, which concluded that the children had normal intelligence often masked by "motor incoordination." But a still larger study concluded, on the basis of standardized tests, that children with FD were significantly retarded in comparison with their siblings. This was a way of constructing the problem that acknowledged the relative affluence and high educational level of the Ashkenazi families in which FD appeared. If children with FD had normal intelligence in families in which most of the children were extraordinarily bright, then clearly FD was affecting their intellectual development (Sak, Smith, and Dancis 1967).

In the study by Welton, Clayson, and Axelrod (1979), fifty-two patients—which at that time constituted one-fourth of all known patients with FD worldwide—were given the Wechsler intelligence test in their own homes "to minimize the predictable emotional stress of the situation." (People with FD have a "tendency to overreact to stress" and a "predisposition to 'give up' in pressure situations.") All the participating families were of relatively high socioeconomic status: forty-three of the subjects had parents who had either college or graduate degrees, and all parents had high school diplomas (709). Yet, despite these generally good family environments, the physical impairment of the FD body, the authors emphasized, created an abnormal learning experience, as a child with chronic illness and impaired motor skills cannot interact with the larger world in the same ways other children do. "Intellectual potential in these subjects is undermined by motor incoordination, sensory deficit, and limited exposure to the object world" (710).

Indeed, the authors blamed any intellectual deficits in children with FD entirely on this physical experience of the injured and impaired body: patients are "clearly more disabled by the emotional and developmental lags caused by their physical impairments than they are by any 'real' diminishment in intellectual potential due to their disease" (710). The "real" in quotation marks seems to refer to genetic diminishment. The authors here suggest that FD does not produce lowered intelligence. Lowered intelligence is instead an epiphenomenon, a side effect rather than a constitutive quality. Their conclusion was that "far more dysautonomic patients are capable of adjusting to standard school programs than was heretofore thought possible" (711).

MOMENTS OF TRUTH IN GENETIC MEDICINE

In a follow-up study apparently involving the same group of patients, the same authors explored personality development by subjecting the patients (again) to standardized tests for ego strength, assertiveness, introversion, frustration level, neuroticism, and so on (Clayson, Welton, and Axelrod 1980). The disease, categorized in terms of skin tests and defective lacrimation, was also accessible through the standardized tests for intelligence, personality, ego strength, motor testing, and maturity. The investigators concluded that people with FD "do not lose their sense of being damaged," even in adulthood, and they reported a "chronic, low depression . . . in many FD adults." They tended to be dependent on their families into adulthood, but there was "no evidence that the incidence of thinking disorders or other 'real' psychotic manifestations is greater among FD subjects than is found in the general population" (274).

In some ways the series of tests and psychological assessments made the FD experience more physical by constructing all mental changes in the patient as the normal response to the bodily state. FD did not produce "real" psychotic experience or "real" diminished intelligence but only a damaged body that the mind found difficult to manage. What effects could be seen as "real" elements of the disease and what were tangential? What was physical and what was mental? Where exactly was FD located? Assessing its location on the genome, in the autonomic nervous system, in the mind, or in the whole functioning body required different skills, technologies, and assumptions. The disease had psychic, genetic, and neurological qualities, and as they interacted these qualities could be confused with each other, the neurological masquerading as mental deficit, for example, or the emotional presenting as a medical crisis in the emergency room.

Familial dysautonomia as a bodily and psychic phenomenon was further obscured by diagnostic uncertainty. Although those who knew the disease could readily see its signs in a new case—in the movement of the body, in the face, on the surface of the tongue—their knowledge was difficult to convey in a scientific paper. Physicians needed a sign that would be present in all circumstances.

A Reliable Diagnosis

Diagnosing familial dysautonomia remains difficult. A skin test, a reaction to intradermal histamine, is now accepted as the preferred technical measure. But those who work with patients with FD, including Axelrod, continue to place considerable faith in the knowledge acquired by contact and inter-action. When Axelrod confronts a possible case of FD, she takes a medical history, asking questions about sweat, tears, reflux—"there is an art of tak-ing an autonomic history." She checks for hallmark blood pressure anom-alies. But in fact she can see the disease. "I can tell by looking at the child." She can see it in the face and in the tongue, which is smooth and lacks fungi-form papillae and taste buds.

Even the administrative staff members of the parent support group re-port an ability to see the disease based on frequent contact with patients with FD. Executive Director Lenore Roseman told me that, on a trip to Russia in 1989, she and fellow staff members visited a man who had "diagnosed this child" with FD, and "we confirmed the diagnosis although we are not doc-tors. We took one look at this girl and knew this was what she had." The in-formed physician, provided with a detailed list of symptoms, might easily miss the disease. But administrators who had interacted with children with FD and their families had bodily knowledge that gave them diagnostic con-fidence.[4] Such knowledge is functionally irrelevant, of course, to a pediatri-cian confronted with a strange suite of symptoms he or she has never en-countered before. Physicians needed a diagnostic test that did not depend on familiar seeing. This was the intended purpose of the intradermal his-tamine test.

Alfred A. Smith and Joseph Dancis described the intradermal histamine test in a 1962 paper in *Pediatrics*, identifying it as a "simple diagnostic test for the disease" that also had "theoretical implications in understanding the mechanism of the disease." In people without dysautonomia, this injection produced a large zone of irritation and a raised wheal at the site of the in-jection. In patients with dysautonomia, the wheal was present but no large zone of irritation, or "flare," was expected.

Their trial of the histamine test on fifty-eight subjects (fourteen of them adults) resulted in a characteristic dysautonomic response in twenty-four of the twenty-seven children who had earlier been diagnosed with FD. The three anomalous responses had to be explained, and Smith and Dancis ex-plained two of them by suggesting that the children did not in fact have FD.[5] The test in effect defined them out of the disease category. The third anom-alous result was in a ten-month-old infant, and this response they attributed to the testing technology. The 1:1,000 dilution of histamine produced a nor-mal flare response in this infant who clearly did have FD, but a dilution of

1:10,000 produced a "typical dysautonomic response" in the affected infant and a normal flare in six other infants who did not have FD. For infants, they suggested, the testing protocol should be different. Later Smith and Dancis elaborated on some of the ambiguities of the skin test, but their general conclusion was that the skin test was the most powerful diagnostic tool available.[6]

By 1970, however, Smith and Dancis reported that the diagnosis had become "a little frayed at the ends." There were clearly other familial disorders that involved aberrations in autonomic function and individuals who seemed to have FD but who did not have the full symptom complex. That the underlying mechanism of FD remained unclear contributed to this diagnostic uncertainty, because as long as the disease was defined solely in terms of clinical symptoms rather than systemic problem, it was difficult to assess which symptoms were "fundamental" signs and which unreliable. The disease was vulnerable to inappropriate poaching by clinicians who encountered patients with any of the diverse symptoms found in children with FD. Two investigators at the National Institutes of Health (NIH), for example, tried to construct a "type II familial dysautonomia" afflicting non-Jewish children with "blond hair, blue eyes and fair complexions" who had diminished sensitivity to pain, unexplained fevers, self-mutilation, and mental retardation. What linked the two conditions, these investigators said, was the lack of taste acuity and the origin of both disorders in the autonomic nervous system (Wolfe and Henkin 1970). But Smith and Dancis (1970) came to the defense of FD and defined these children as outside the range of FD proper. The problems of these children seemed to involve "fundamental differences in pathophysiology." The term *familial dysautonomia* furthermore encompassed "a group of children now sharply defined by clinical history, genetics and a series of objective criteria." The condition described by Wolfe and Henkin (the NIH investigators), they said, should be called congenital sensory neuropathy—the term "preferred by previous authors."

Naming the disease and deciding when it was present was a way of bounding the condition not only in terms of its bodily manifestations (which it might share with other conditions) but also in terms of its social and ethnic limits. Those engaged in establishing a reliable diagnostic test needed to fend off claims that other bodily conformations should be subsumed under FD, which already had both an identity and a constituency. Indeed, in the 1970s and 1980s, many other hereditary sensory and autonomic neuropathies (collectively known as HSANs) began to be described, so many that FD is now classified as the most common of the HSANs. "Although all the HSANs are generally characterized by widespread sensory dysfunction and variable autonomic dysfunction, they can be distinguished by specific and consistent neuropathologic lesions" (Axelrod 1997b, 525).

Most physicians who encountered an infant with FD would presumably be unfamiliar with the rather esoteric debates about the condition's precise boundaries and its appropriate signs in the newborn. The disease was rare and not likely to come to the attention of a pediatrician who had never heard of it, much less seen it, and who had no knowledge of the characteristic lesions of the FD body. It was this professional population that the physicians working with FD needed to reach.

In a 1979 paper on neonatal diagnosis of FD, a group of physicians at Hadassah-Hebrew University Medical School sought to describe "the clinical picture in three neonates" and offered "guidelines beyond the usual reference to 'feeding difficulties' and 'floppiness.'" There was no "clinical sine qua non" for FD, they suggested, and even "important features such as Ashkenazi Jewish ancestry and altered catecholamines in the urine are not invariable." The disease was also variable in its expression—one of the three cases they described was almost asymptomatic, with no gastrointestinal symptoms, no tendency to pulmonary aspiration, no abnormal behavior, and an initial equivocal histamine test. FD was finally diagnosed in this infant at age six months because of lack of tears and repeated positive histamine tests. In addition, some of the features of FD in the neonate might be considered normal in premature infants, including skin mottling, absent tearing, and limited reaction to pain. "There is a natural reluctance to make a commitment about clinical findings such as absence of fungiform papillae of the tongue when so much depends on the diagnosis" (Perlman, Benady, and Saggi 1979).

The confusing picture of FD in the neonate that this 1979 article presented seemed to suggest that diagnosis would be impossible without familiar seeing. Somewhat ironically, the authors, after having taken apart the reliability of virtually every sign of FD in the neonate, proposed that the purpose of their paper was to "reinforce the confidence of the pediatrician confronted with the responsibility of diagnosing FD in a neonate" (241).

In practice, in the 1970s and 1980s, many patients with sensory disorders found their way to the one person whose familiar seeing could be trusted, Felicia Axelrod.

Felicia Axelrod and Clinical Management

Since the 1970s, clinical management of familial dysautonomia has had a profound impact on the lives of those affected. People with FD survive longer and function better when they are enfolded into the system of surgery and drugs developed by one physician, Felicia Axelrod. If the development of a genetic test for FD dreams of a clean, straightforward solution, Axelrod's project is intensely practical reality. She concerns herself with aspiration, tongue sores, drooling, spinal curvature, constipation, eyelid infections, and spells of weakness, and she advises parents and patients on how to handle schooling, toilet training, emotional maturation, social life, swimming lessons, and airplane travel.[7] Axelrod's world is not the world of gene sequencers or restriction fragment length polymorphisms or Southern blots but of gastric secretions, low-grade infection of the eyelids, and inhalation therapy.

Genetic diseases sometimes appear in popular culture and other sources as treatable through an imagined (future) gene therapy or through prevention strategies dependent on fetal testing. But in practice, in most cases, genetic disease is treated with demanding regimens of aggressive clinical care. For those who must help a child with FD through an average week of classes, homework, play dates, and errands, neurons and chromosomes are remote, abstract entities. Axelrod's *Manual of Comprehensive Care* (1997a) is a crucial resource. "Based on knowledge and experience accumulated from dysautonomic children and their parents," the manual provides a glimpse of what one parent of a child with FD has called "mothering on Mars" (Ginsburg and Rapp 1999).

Felicia Axelrod's involvement with FD began in the late 1960s. She attended New York University (NYU) medical school and in 1966 completed her residency at NYU's Bellevue Medical Center. Joseph Dancis, also at NYU and the co-developer of the intradermal histamine test for FD, then asked Axelrod to help him with some hospitalized children with FD. Axelrod became interested in these children, partly because of what parents told her. She tended to see such children only when they were in crisis, but "parents said the patients were so different when they were well,"[8] and she found this to be true. "When the crisis is over they are so different and so wonderful that you want them to enjoy that part of their life more."[9] She began observing and cataloguing patients with FD with the goal of providing "continuity" in clinical care. By the end of her first year of work with FD, she was seeing and tracking the progress of fifty-four patients.

In 1970, with the support of the Dysautonomia Foundation and of NYU Medical Center, Axelrod established the Dysautonomia Treatment and Evaluation Center at NYU. She coordinated care of the patients, arranged for them to see specialists in New York, and created a clinical record of successful and unsuccessful therapies. Basically Axelrod adopted an empirical treatment policy, not discursively constructed as scientific research but constructed as classic medical trials based on the individual judgment of the physician. Whatever worked well with one patient she tried with others. Axelrod gradually came to have clinical control over virtually every child in the world who had FD.

In the process, she began working with various technologies, both surgical and chemical, for the management of the disease. The most important such technologies were arguably those that permitted the children to consume adequate nutrition. Patients with FD find it difficult to swallow and they have consistent problems with vomiting. These problems can lead to pneumonia as liquids reach the lungs and can result in poor nutrition, which can have profound consequences for mental and physical development in young children. Eating, for the child with FD, is a difficult challenge. Axelrod's solution, now almost universally applied, was to attack the eating and vomiting problems with surgery.

In 1978 she began treating children with gastrostomy, the surgical creation of an opening into the stomach through the abdominal wall to allow tube feeding, and fundoplication, the tightening of the lower end of the esophagus making it more difficult for vomiting to occur. There were immediate improvements in weight in these children, and in 1985 the value of the two surgical procedures, both for nutrition and for the prevention of lung infections, was statistically demonstrated. After 1985 the surgeries were performed even earlier in the life of the child, and by 1993, 70 percent of children with FD under age five had gastrostomies.[10]

The gastrostomy-fundoplication strategy is characteristic of Axelrod's approach to the management of FD: aggressive, intense intervention in symptoms, with the expectation that this intervention will produce something resembling normal life. According to Axelrod's care manual (1997a), the gastrostomy should be explained to friends at school, if they should inquire, by the child telling them, "I have trouble swallowing and this is how I get a

MOMENTS OF TRUTH IN GENETIC MEDICINE

drink." And parents should handle the tube feeding "in a positive way" so that the child accepts it. "It is imperative that a normal family life be maintained," Axelrod advises. "Parents should be encouraged to pursue their own goals." Over and over again, in her comprehensive care manual and in interactions with parents, Axelrod constructs the child with FD as capable of almost all normal childhood activities and encourages parents to treat the children with a minimum of fuss. Manage the body medically with dramatic surgical and chemical interventions, she suggests, and then manage the body socially by building around it something that can be seen as normal life.

This idea has its poignant contradictions. Normal life is a product of surgery, tube feeding, Valium, supplemental oxygen on airplane flights, suctioning of chest secretions, and anticonvulsants. Axelrod's care manual expresses a profound longing that medical care and parental diligence can compensate for occult malfunctions and that clinical strategies can standardize the FD body.

Axelrod's Dysautonomia Treatment and Evaluation Center now evaluates patients with FD annually, producing an "evaluation letter with specific individualized recommendations" for the primary care physician.[11] Despite the difficulties of air travel for these children, many parents fly to New York annually or semiannually to bring their children to see Axelrod. Many also come to New York each June for Dysautonomia Day, when scientific findings are presented for parents and social support meetings are held. And although there is a satellite center for FD in Israel, and scientists in many countries are working on FD or related HSANs, Axelrod is the clinical leader in the field, the single most important person in the transformation of the disease. She has tremendous power over the lives of these families and is therefore the focus of both adulation and occasional rage in FD circles.

Axelrod is also a leading proponent of the prevention of FD. The center at NYU sees about fifteen new patients a year, and Axelrod has made it clear that she wants to see that number fall. She would like to eliminate the disease to which she has devoted her professional life—the disease around which her center has been built, and the disease for which she is the world's leading expert. She foresees the dismantling of the intellectual and social empire that has consumed so much of her energy, at some future time when children with FD are no longer born. "It is in fact quite possible that by 1999, which will be the fiftieth birthday of the first description of FD, we will have reached a point that we will no longer have any new cases of FD being born, and that all affected individuals will be able to look forward to a more hopeful future as we enter the era of gene therapy," Axelrod said at the 1994 Dysautonomia Day. "I look forward to conquering FD by 1999."[12] The site of this future conquest, for FD as for so many genetic diseases, is presumed to be the genome.

Finding the Gene

The Dysautonomia Day conferences held annually since 1986 in New York City are socioscientific events, featuring awards to successful young patients, reports on the latest technical data, parties, dinners, and play dates. Axelrod gives a talk. The foundation staff presents reports. And the assembled families hear about the search for "George."

"George" is the name given to the FD gene by the Massachusetts General Hospital project leader, Susan Slaugenhaupt. In narrative presentations apparently intended to simplify and explain molecular genetics to a nonscientific audience, Slaugenhaupt presented the FD gene as a rakish man in a smoking jacket, with a cigarette; an unreliable worker, sleeping on the job. "What can we do to wake him up?" Axelrod asked in her 1997 Dysautonomia Day report.[13] Carrying the analogy further, Slaugenhaupt proposed that the lab was searching for George's foot in an apartment complex in the Bronx or Queens.[14] The images of rakish bon vivant, of George's foot, and of a dragnet through New York boroughs were translations of the technical into the social. Whatever they might know about the experience of the disease, parents attending a Dysautonomia Day celebration might not know anything about gene sequencing. Just as their emotions (enervation, say) could be translated into data, appearing in the passive voice in a scientific paper, so technical details could take on the emotional tenor of a crime story, a search for someone devious and dangerous through the streets of New York.

Until 1997, the work at Harvard and Massachusetts General Hospital by Gusella, Slaugenhaupt, and their team was funded by the Dysautonomia Foundation. But in 1997, the National Institutes of Health approved a three-year grant of $190,000 per year to support the FD research. Parents, seeking technical knowledge, gave the Harvard group the intellectual resources to seek and acquire NIH funding. "Finally, your tax dollars are helping to fund the FD research," Slaugenhaupt told the parents.[15] But George was difficult to find, even with NIH funding. By 1999, the entire candidate region of DNA had been mapped, 471,000 base pairs, and five candidate genes had

MOMENTS OF TRUTH IN GENETIC MEDICINE

been identified, but there was still no gene. Slaugenhaupt promised the parents that the Harvard group would "keep sequencing until we find the mutation, aka George."[16]

Increasingly in the 1980s and 1990s, genetic support groups began to define their mission as contributing not only information and insights but also, like the FD group, funding for scientific research. The Fanconi Anemia Research Fund, Inc., for example, is a parent support group named to reflect its emphasis on fund-raising and scientific research. The group began after an Oregon family learned in 1983 that their oldest child suffered from Fanconi anemia. In 1985, feeling "isolated and alone," the parents decided to form a support group, working with a researcher who had a Fanconi anemia registry and was willing to forward a letter to her families about the proposed group. That initial mailing produced nineteen responses, and these families became the core audience for the newsletter, special news bulletins, and the handbook for parents of children with Fanconi anemia. By 1994 the group had 350 families from forty-six states and twenty-one foreign countries.[17]

In 1989 the Fanconi anemia group decided it "needed to become a 501(c)3 organization"—one that could raise funds and provide grants to scientific researchers. "We needed to become a legal entity to raise and distribute funds." The group appointed a board of directors that included three family members of patients with Fanconi anemia, as well as scientific representatives and health care professionals. Within five years, the group had given fifteen grants to researchers studying Fanconi anemia, most engaged in efforts to isolate the gene rather than in studies of therapies or patient management. The organization has also sponsored meetings of parents and of researchers, and in 1990 sponsored the creation of a Fanconi Anemia Cell Repository at Oregon Health Sciences University in Portland. Families contribute blood and skin samples to this repository; researchers can request such samples if they are working on Fanconi anemia (J. O. Weiss and Mackta 1996, 133–36).

The families and patients thus become the advocates for work on the disease. They seek to provide funding, to establish scientific resources, and to bring together scientific researchers. They also provide tissues and blood, the physical materials that are central to the research. They are part of the infrastructure that shapes the public status of genetic disease and research in molecular genetics.

More recently, some individual parents have become named contributors to scientific papers or even genomic entrepreneurs. They enter the technical system as a result of their experiences as parents and in the process are legitimated as technical experts. The anthropologist Karen-Sue Taussig has explored, for example, the active participation of Patrick and Sharon Terry in the research community involved with the disease that their children have, pseudoxanthoma elasticum (Taussig 2005). When the children were diagnosed in 1994, Patrick was managing a construction company. He had attended community college for two years. Sharon was home-schooling their son and daughter. She had a master's degree in religious studies. Neither, therefore, had the credentials usually understood to be necessary for technical knowledge production. But in 2000, Sharon Terry was a coauthor on two papers in *Nature Genetics* that announced the discovery of the gene for pseudoxanthoma elasticum (Bergen et al. 2000; Le Saux et al. 2001). And Patrick Terry joined Randy Scott, the founder and former chief executive of Incyte (a major biotech company), and three others in securing $70 million in venture capital to establish Genomic Health, a new biotechnology firm. While the Terrys are unusual for their very high level of technical and economic involvement in genomic medicine, they are only slightly so. Many families affected by genetic disease are deeply engaged with the scientific community (Taussig, Rapp, and Heath 2002).

The private gene search of the Dysautonomia Foundation that began in 1990 was "unprecedented" in the sense that virtually all other gene hunting had been funded by federal agencies or private biotechnology companies. The FD gene was not a gene whose study anyone thought the NIH would wish to fund. When a foundation board member became frustrated with the research the foundation was supporting, he spoke with scientists about the possibility of finding the gene and convinced his fellow board members that genetics research would be more productive. The group asked five leading geneticists whether they would be willing to search for the FD gene. Only James Gusella at Harvard University was interested. The foundation began providing him with about $400,000 per year. In 1994 he found markers for the gene on a section of chromosome 9, making possible a prenatal test in affected families.[18] These markers, which could be used to screen pregnancies in families with one child already affected, were credited with making possible "23 healthy babies born all unaffected with FD as predicted."[19] The foundation did not, in this document, report how many fetal tests were pos-

MOMENTS OF TRUTH IN GENETIC MEDICINE

itive and therefore led to abortion of affected fetuses. The fetus that has FD is not an object of public mourning, while one that is free of FD is an object of public celebration.

Gusella and his group at the Harvard Institute of Human Genetics later applied to the NIH for three years of funding, at $190,000 per year, to continue to pursue the gene, and this application was successful. In their NIH application, they said that their proposal "included all the work we have done on FD to date, including the initial collection of families."[20] But the group at Harvard did not collect the families. The families, as a social and medical conglomerate, collected Harvard. Emotional knowledge does not just provide comfort to those in pain. It also produces scientific papers and gene maps.

In January 2001, two different scientific groups announced that they had identified the precise gene causing familial dysautonomia. "Race to discover gene mutation ends in virtual tie," said one press report. "In a remarkable confluence of scientific sleuthing, two medical teams, working independently, discovered the gene mutation that causes Familial Dysautonomia within days of each other" (Tauber 2001). The teams, based at Fordham University in New York and Massachusetts General Hospital in Boston, were both to publish reports in the March issue of the *American Journal of Human Genetics*. A newcomer, Berish Rubin at Fordham University in New York City, drew on the extensive published data of the Harvard laboratory to identify the gene, provoking the Harvard group to publish further data, also identifying the gene, earlier than planned. The story of this "remarkable confluence of scientific sleuthing" provides insight into the structure of contemporary genomic science (S. L. Anderson et al. 2001).

Spring 2001

In the spring of 2001, as news about the discovery of the FD gene by two different laboratories unfolded in scientific announcements, in the mainstream press, and on FD websites, the fragile alliances binding the support group, the treatment center at NYU, the parents, and the Harvard research group began to fall apart. A new "splinter" parent support group was formed. The two labs, publicly cordial, were competing both for control of the patent for the genetic test for carriers and for the loyalty of the parents. And the two parent support groups were offering different visions of the needs of this specialized community. Rumors, accusations, and angry messages appeared in email and on the parent websites. For some parents, the

Harvard group seemed to have betrayed their trust by delaying publication of the results. For other parents, the Fordham group's motives were problematic, as Fordham used the technical work already completed by Gusella and his group at Harvard to solve the final problem.

The papers in the *American Journal of Human Genetics* therefore told only a small part of the story. Finding the FD gene was a social event, a political event, and not just a technical event. A truth about nature, that FD was caused by a mutation in the gene encoding the IKAP protein, was embedded in rage, desire, betrayal, profit, and hope.

The group at Harvard and Massachusetts General Hospital had, as noted earlier, been sought out by the parent support group in 1992. Other geneticists had turned down the offer of funding to search for the FD gene, but Gusella, a skilled gene hunter, was both confident that the FD gene could be found and interested in trying. His group quickly developed a test that could detect the likelihood of FD in a fetus prenatally in families in which one child with FD had already been born. This was the kind of test that depended on a genetic marker commonly inherited with the condition but not implicated in causing it. The group localized the gene to chromosome 9, and from 1993 to 2000 they published four papers on their research, including a paper in *Mammalian Genome* with detailed and specific information about the location and function of the gene (Chadwick et al. 2000). Meanwhile Gusella's group filed for a patent for the genetic test. Every June, Slaugenhaupt participated in Dysautonomia Day to describe the search for George. And by about 2000, some of the parents hearing these presentations were becoming discouraged at the slow progress the group seemed to be making. Later, when the competing lead researcher, Berish Rubin, described his motivation for seeking the gene, he echoed this perspective, saying that he decided to look for the gene because he was discouraged by the limited progress being made at the Boston laboratory.

Rubin was certainly an unlikely gene hunter. In the biology department at Fordham, Rubin had published work on interferons and AIDS. He was not an accomplished genomics scientist like Gusella. He had no track record of finding genes. But he had another form of expertise, a form that became important as events unfolded. Rubin had FD in his family. One of his nephews had a child with FD, and then other siblings in the same family had children with the same disorder. Rubin had a personal stake in the disease, and he came into the community first as a family member and then

MOMENTS OF TRUTH IN GENETIC MEDICINE

as a newly realized genomics mapper. According to Rubin, in 1999 he was asked by Dor Yeshorim, an orthodox Jewish group, to work on finding the gene, but at that time he told journalists that he thought Gusella's group was doing a good job and that he should not pursue the invitation. A year later, however, he decided that "there really wasn't much progress under way" and "I thought I would give it a shot" (quoted in Tauber 2001). In the spring of 2000, Rubin's FD project began. Just a few months later, in December, the group had found the gene, an outcome Rubin attributed to the singular focus of his team.

Technically Rubin was first, announcing on 9 January 2001 that he had found the gene. But the Massachusetts General group had apparently already found the gene but hesitated to announce it; on 12 January, the Boston group said it had the gene as well.

For the parents, these twin announcements were a mixed blessing. The lab that the established parent support group, the Dysautonomia Foundation, had supported had officially been scooped. And someone clearly had provided Rubin with cell lines and patient histories, outside the regular channels. Slaugenhaupt and her team had possibly been betrayed by someone in the FD community. And Slaugenhaupt and her team had possibly withheld publication in the hope of asserting greater control over the patent, a withholding that some interpreted as a betrayal of the support group. To add to the parents' troubles, the mechanism in the gene was confusing and unusual. Some of the frustration expressed in the ensuing internet discussion resulted from this confusion about the gene itself, and parents exchanged long messages intended to decipher just what it meant for them and their families that the gene was so unusual.

Familial dysautonomia is caused by a mutation that produces a splicing error in affected tissue. Two single base-pair mutations are responsible in nearly 100 percent of those affected. The mutation results in abnormal splicing in the RNA, such that one of the coding regions for the IKAP protein is lost. So the FD gene causes a glitch in the cellular mechanism that has huge systemic effects. A piece of the protein drops out, and the autonomic nervous system malfunctions.

In February 2002 some of the parents formed a new support group, based in Chicago. FD Hope was created as a public, tax-exempt, nonprofit organization that would support scientific research on FD and provide the same kinds of social support services offered by the older Dysautonomia

Foundation. Its connections to the treatment center at NYU were cordial but more tenuous, and its research agenda was focused more on treatment than on genetic testing. Genetic testing, of course, does not benefit children who have FD. It makes it possible to eliminate the future birth of such children, but it does not play any role in the day-to-day management of the disease.

On 26 June 2002, Genzyme Genetics announced that FD would be included in an expanded menu of carrier tests offered to Ashkenazi Jews. This was the first commercial marketing of the FD gene. The expanded test menu included nine fatal or debilitating genetic diseases found in Ashkenazi Jews. A single blood sample taken from both members of a couple could be used to determine carrier status for all nine disorders, all of which are autosomal recessive genes.

Conclusion

In a pamphlet intended for families with genetic disease, the president of the Alliance of Genetic Support Groups proposed that genetic services by the federal government had their origins in the creation of the U.S. Children's Bureau in 1912. In this historical construction, genetic diseases are a natural focus of federal interest in the health of mothers and children. The appropriate founding agency is the U.S. Children's Bureau, the same bureau that organized the first field trials of the phenylketonuria test (see chapter 2), which was the first sustained program in public health genetics. This alliance official thus linked the omnibus Genetic Diseases Act of 1976 to the child labor issues that originally shaped the creation of the Children's Bureau (Mackta 1992).

In this same pamphlet, the author proposed that "consumers" of genetics services are experts in their diseases. "A consumer's life experience and knowledge are essential to making good use of the scientific information and clinical experience of the professional." The victim of genetic disease thus becomes one of those who can shape its interpretation. Parents of children who have genetic disease, she proposes, can contribute information to professionals, helping them understand both practical and theoretical questions. Consumers are a "valuable resource" for health care professionals. They contribute to the meaning of genetic disease.

In their scholarly duet on fetal images, anthropologists Faye Ginsburg and Rayna Rapp explore the segregation of knowledge of disability from medical practice and from feminist theory, two realms in which the fetus is an important actor. Rapid technological changes in prenatal testing and fetal imagery have converged with rapid social changes in the legal and cultural meaning of disability to make prenatal testing an extraordinarily complicated experience, they suggest. The fetus is situated precisely at the intersection of reproductive politics and disability rights, and this freighted cultural location makes informed consent—"socially informed consent"—an elusive goal. "Most women in industrialized countries sent for prenatal testing have little or no personal, social or political knowledge of the disabilities for which their fetuses will be screened and about which they may have to make decisions." Without such knowledge, assessing the meaning of any particular disability is extremely difficult. Medical knowledge of disease and social knowledge derived from living with disability are effectively segregated, and pregnant women make decisions about reproductive tests in isolation from the "ghettoized experiences of families raising disabled children" (Ginsburg and Rapp 1999, 279). For Ginsburg and Rapp this work has a direct personal dimension that informs their essay. Ginsburg's daughter Samantha has familial dysautonomia; Rapp had a positive prenatal test for Down syndrome some years ago and chose to abort. They are consumers of technical knowledge who have also acquired emotional knowledge.

The physical location of knowledge (in what bodies does it reside?) depends on a political process. Labor that seems to be entirely mental and entirely situated within the scientist becomes, on closer examination, physical, material, emotional, and widely dispersed across a social network. This is certainly the case with FD, which has always been a collaborative project.

In June 2003, Berish Rubin produced another end run around the Dysautonomia Foundation, proposing a possible therapy that could be purchased in a health food store. Rubin announced that a form of vitamin E supplement, called tocotrienol, had been shown to increase cellular levels of the IKAP protein, deficiencies of which are believed to be the primary cause of the disabilities produced by FD. The details were in a scientific paper, "Tocotrienols Induce IKBKAP Expression: A Possible Therapy for Familial Dysautonomia." But evidence also came from parents of six children with FD, who were told to give their children 50 mg of the over-the-counter supplement daily for "a while" to see what would happen. Rubin asked parents and children to report their experiences to him. "We did not know what to expect, but with the availability of tocotrienols as an over-the-counter vitamin supplement, we thought it was worth a try. I have to tell you that the results observed have surprised me," Rubin wrote in an online message to the FD community. The children taking the supplement had increased energy, weight gain, fewer crises, no crises with fever, and an increase in the amount of tears produced. His message went on to describe how to manage the oily, sticky product, sold in a gel cap but presumably difficult for a child with FD to consume in that form. Parents should pierce the gel cap and squeeze the oil onto food, which should be consumed immediately as exposure to air destroys the supplement (www.familialdysautonomia.org/News.htm).

The new therapy, then, was a low-technology solution (a health food supplement) that was expected to be assessed based on familial expertise and folk epidemiology. FD had been embedded in technical domains at two levels (the gene and the phenotype), managed by mapping and by surgery and drugs. Rubin's vitamin E strategy embedded FD in a third domain, folk knowledge.

As I have shown here, over the past fifty years different kinds of knowledge about the FD body and its properties were generated by parents, physicians, the staff of the parent support group, and the patients themselves. Even if the knowledge of parents is coded as outside the boundaries of biomedicine, it has been procedurally there from the first diagnoses of FD in the 1940s to the search for the gene in the 1990s, from the first effort to extract information from parents bringing children to Babies Hospital to the most recent work in James Gusella's lab. The active and aggressive presence of patient and parent support groups is one striking continuity in the practices and culture of human genetics over the past few decades. Families provide emotional support and practical information to each other but also provide blood samples and data, and sometimes they end up inside the lab or even recalibrate their scientific expertise to bring it to bear on a phenomenon with which they have familial experience. Every scientific publication on FD I have encountered includes a note of thanks to the Dysautonomia Foundation for providing contacts and referrals. The foundation has functioned at one level as a tracking system that organized particular bodies, selected those rare people in whom the disease was manifest, and established a database that simplified the scientific study of the disease.

Familial dysautonomia is a classic genetic disease (recessive, strange, rare, of limited importance in everyday clinical medicine) that has been transformed at virtually every level in ways that reflect the broader transformation of genetic disease after 1945. FD moved from a cluster of difficult

MOMENTS OF TRUTH IN GENETIC MEDICINE

cases in a single New York hospital to an international network of websites and genetic testing, and from a grueling ordeal of patient management and early death to a condition preventable, at least in theory, by prenatal diagnosis and selective abortion. The disease was modernized through a system of medical diagnosis and analysis, parental involvement, scientific research, communications technologies, and technologies for the aggressive daily management of the disease. Children with FD now commonly attend school, and about half of them survive both the disease and its attendant medical care into adulthood.

Despite the many changes in the technical, medical, and social meaning of FD, at the level of the individual affected it remains a highly demanding bodily state. "Autonomic dysfunction causes almost every other system to dysfunction in almost a domino kind of fashion."[21] Children with FD are subject to feeding, breathing, skin, coordination, and bone problems. They can experience a "dysautonomia crisis" in response to stress, fatigue, or infection. And they can pass out on airplane flights when their bodies fail to adjust to the reduced oxygen in a pressurized cabin. They often have diminished pain sensation on some parts of their bodies and can be burned or injured without realizing it. They may find it difficult to eat and swallow. For reasons that remain unclear, the disease produces a "developmental arrest" in the autonomic nervous system (Axelrod 1997b, 528). As a consequence, individuals with FD move through the world differently; they are commonly breech births (Axelrod, Leistner, and Porges 1974) because "these babies don't know where they are,"[22] and their physical disorientation can persist into adult life.

Between 1970 and 2004, 595 people were diagnosed with FD; about 340 patients were still alive in 2004, and about 35 percent of these patients were more than twenty years old, a sign of how much the clinical interventions developed by Axelrod have enhanced survival in this very serious disease (www.familialdysautonomia.org). Twenty-eight had married, and five had unaffected children of their own. All these patients are catalogued in the files of the Dysautonomia Foundation, and all, either directly or indirectly, are under the care of Felicia Axelrod. The FD diagnostic and treatment regimen is well-established, the institutional structures are in place to increase physician awareness and produce more rapid diagnoses, and efforts to improve treatment continue. Yet the dream of almost everyone involved is that this infrastructure of technology, knowledge, and practice will cease to be necessary. Those who know FD well do not want to see it again. One staff member at the foundation cries whenever a baby is newly diagnosed with FD.[23]

Here, too, FD mirrors much larger trends. Genetic diseases are usually complex and difficult to treat, and the most efficient way of managing them is by early fetal diagnosis and abortion. Preventing the disease from ever occurring, with the help of a gene or genetic markers, is technologically simpler than developing a full understanding of the accompanying physiological deficits and treating the fundamental, underlying problem—which for FD, as for so many genetic diseases, is still not fully understood.

As a disease of Ashkenazi Jews, FD furthermore provides insight into the ways that ethnic identity folds into definitions of disease. Babies with no known Ashkenazi background who present with the standard symptoms of FD violate medical expectations and are often diagnosed out of the category. And scientists resisted proposals to link similar diseases that were not found in Ashkenazi Jews to FD, partly on the grounds that FD already had a strong ethnic identity. The specificity of the disease was enhanced by its relationship to a recognized social and religious group, and the symptoms and physiology alone did not define the relationship of this disease to other diseases. The sorts of bodies in which it could be found defined it as well.

In late 1997, publicity about a new mutation linked to colon cancer in Ashkenazi Jews led Jewish leaders to express concerns about stigmatization. Jewish diseases were already the subject of high-profile public health programs, such as Tay-Sachs disease, and of textbooks entirely focused on the genetic burdens of this small population. The breast cancer and colon cancer genes were both identified as linked to Ashkenazi Jews, and the director of American affairs for Hadassah noted that "this feels uncomfortable to the Jewish community." But, she said, "we recognize how important this lifesaving research is."[24]

The changed meaning of genetic disease, its expansion, and its burgeoning preventability has been played out even on stages with just a few players, through many of the same processes that have shaped the cultural and medical meanings of such genetic diseases as cancer, heart disease, and (most recently) old age. FD acquired preventability by virtue of parental desire, and it became a part of a linked network of advocacy groups devoted to the treatment and prevention of genetic disease. This network incorporated the knowledge acquired by caring labor, knowledge that could lead to the peripheral neurons or to chromosome 9. Both the emergence and the

MOMENTS OF TRUTH IN GENETIC MEDICINE

(impending) decline of FD are technological phenomena: it was invisible before antibiotics, and genomics may make it invisible again.

In January 2002 the NYU anthropologist and MacArthur Award recipient Faye Ginsburg appeared in a full-page photo in *Vogue,* opposite an advertisement for Le Rouge makeup. Ginsburg was smiling, glamorous, dressed like a fashionable intellectual and made up like a model. She was pictured in *Vogue* because she is a carrier of FD, as is her husband Fred Myers, who is also an anthropologist at NYU. Their daughter Samantha Myers has become a spokesperson for FD. The photo in *Vogue* was pegged to a promotion by a makeup manufacturer that was supporting the Make-a-Wish Foundation. Samantha had earlier been in the children's television program *Nickelodeon News* to discuss FD. It was the wish she chose when she appealed to Make-A-Wish: instead of a trip to Disney World, Samantha wanted the opportunity to tell other children about her life so that they could understand her disease. "Samantha gave a personal account of her disease, hoping not only to reach children similar to her, but doctors as well," said the text accompanying Ginsburg's photo. This text also announced that Faye and Samantha were celebrating the discovery of the FD gene, the establishment of genetic testing, and the funding of new research that could lead to better treatment.

Samantha and her parents and family, and many other FD families, support research relating to the gene and research for improved care and treatment. They publicize the existence of the disease in the hope that new babies with FD will be diagnosed earlier in life. And they pose for fashion shoots or appear on television or write books about living with the disease (Arnstein 2000). They run websites, they debate research strategies. About half of their children live to age thirty, and some of them live much longer, and some of them work and have families themselves. Women with FD, if carefully managed through pregnancy, can bear children who are free of the disease, though all will be carriers. Together, these families occupy a brief window of opportunity and contradiction, a moment of historical convergence, in which FD is both survivable and preventable. Its visibility in the body, the genome, the pedigree, and the media is literally an act of human creation. It is a poignant demonstration of the interrelatedness of nature and culture, and the social and the technical.

CHAPTER 7

[CONCLUSIONS]

Except for being hit by a car while crossing the street,
all disease is genetic.
—"The Gene Doctors," *Newsweek*, February 1984

In 1994, cancer researcher Bert Vogelstein told a reporter for the *Journal of the American Medical Association* that "in the last 10 years and especially in the last 5 years, there has been a revolution in cancer research. I can sum up the revolution in one statement: Cancer is in essence a genetic disease" (Breo 1994). Vogelstein proceeded to describe a mechanical system of active oncogenes and damaged tumor-suppressor genes and proposed that "we can now say that [the] accumulation of mutations in several oncogenes and/or suppressor genes causes cancer, just as polio is caused by the polio virus and AIDS is caused by the human immunodeficiency virus." Damaged genes, in this construction, have an independent causal force. Mutations happen because cells are "designed to make mistakes when replicating," Vogelstein proposed, but a single mutation could be the first step toward cancer. A natural event, designed into the cell (mutation), could predispose a cell to become cancerous. At the same time, cells are constantly exposed to environmental forces that encourage mistakes. "The ratio of environmentally-caused mutations vs. replicating mistakes is unknown," he said. (But how exactly to tell the difference?)

Also in 1994, eighty-seven patent citations were filed for types of gene therapy focusing on cancer. Among the many patent assignees were the Salk Institute of Biological Studies, Yale University, Hoechst, Rhone-Poulenc-Rorer, and the Dana-Farber Cancer Institute—in other words, the academic and commercial institutions that process contemporary biomedicine.[1]

Cancer was the first disease for which human gene therapy was tested and applied, and in the spring and summer of 1990, human gene therapy for cancer enjoyed a shimmering public spotlight. Gene therapy in general was then still viewed as extremely promising, and many other methods of controlling cancer were viewed with increasing skepticism and disappointment. The first human gene transfer effort was a tumor-infiltrating lymphocyte marking study led by Steven Rosenberg of the National Cancer Institute. The study had been approved by the National Institutes of Health (NIH) Recombinant DNA Advisory Committee and by the director of the NIH in 1989. It was intended to treat melanoma. But, like every other gene therapy trial that followed over the next fourteen years, it did not accomplish the cure that researchers anticipated. In 2003 the U.S. Food and Drug Administration (FDA) had not yet authorized any gene therapy agent as safe and effective in the treatment of cancer (M. Gottesman 2003).

When Rosenberg's work was just beginning, the then-director of the National Cancer Institute, Samuel Broder, characterized the new gene therapy trial as important as "a symbol of clinical applications of investigator-initiated research" and because it "captures the imagination of the public and in that sense strengthens all of our activities." Gene therapy for cancer might not only lead to better treatment of cancer patients but also increase public interest in cancer research. W. French Anderson told the President's Cancer Panel that genes could be safely delivered to cells and made to work. He complained about the "regulatory fortress" that had to be surmounted to gain approval "the first time that foreign genes were ever inserted into a human."[2]

In 2003 Michael Gottesman recalled the early enthusiasm and suggested that it was still justified, despite the lack of actual clinical success. "There has been disappointment that some of the targets are not as easily inactivated as theory would predict, and the actual delivery of gene therapy in a completely safe and efficient manner has turned out to be a very complex task. Nonetheless, it remains clear that the rationale behind cancer gene therapy is still as strong as it was 14 years ago" (501).

Francis Collins, the director of the National Human Genome Research Institute, wrote in 1998 that cancer genetics was "the Starship Enterprise of the new genetic medicine." His comments bear quotation at length, for they suggest just how important cancer has been to genomic medicine:

> In all of medicine [cancer] is likely to be the field where the molecular understanding of a common and often lethal disease advances most rapidly, where the use of widespread predictive testing to reduce further illness will get its first major application, where most health care providers will first come to grips with the need to incorporate genetics into their practice, where battles about genetic discrimination will be won or lost, and where the compelling paradigm of molecular medicine, that gene discovery will lead to better therapies, will first be put to a real test (xiv).

Collins, like James Watson before him, recognized the centrality of cancer to health care and culture in the United States. The contemporary technical literature on cancer is vast. Dozens of specialized journals have *cancer* in their title. Scientists, physicians, health care workers, biotechnology executives, patients, social scientists, and others publish accounts of cancer, its diagnosis, treatment, social consequences, causes, and political status. At a less technical level, a search of the web turns up thousands of places where information or misinformation can be acquired about many kinds of cancers, risks, or alternative therapies. Rumors about commonplace objects or substances (bras, deodorant) that supposedly cause cancer abound. Chain letters by email exploit reactions to children dying of cancer. Spiritual websites offer solace to families and patients. Cancer, not a single disease but hundreds of diseases, has attracted virtually all the explanatory resources available in industrialized society.

In the context of my study of the rise of genetic disease to medical prominence, cancer provides a vivid example of a condition identified as genetic but understood to be genetic only within an overlay of ambiguity, uncertainty, and environmental contribution. What does it mean to say cancer is a genetic disease? It means to select one thread of causation and place it at the center of the story. It can mean that cancer cells have disordered chromosomes. It can mean that sometimes the nature of the damage is diagnostic. It can mean that there is a higher risk of cancer in some individuals because they are genetically vulnerable. Barton Childs has noted that cancer is "different, perhaps unique, in that the proximate cause is sometimes hereditary, sometimes derived from somatic mutation, and sometimes both" (Childs 1999, 222).

When a widely cited Scandinavian study in 2000 suggested that environment was much more important than genetics in all cancers (Lichtenstein et al. 2000), "a blizzard of press accounts emphasized that inherited genes play only a limited role in the causation of cancer." Because the study was published soon after the announcement that scientists were rapidly nearing their completion of the map of the human genome, some journalists suggested that the attention devoted to both the Human Genome Project and the study of cancer-associated genes might be misplaced. But Joseph Fraumeni, the National Cancer Institute scientist who helped identify Li-Fraumeni syndrome in 1969 and who continued to play a crucial role in molecular epidemiology, responded to these concerns by pointing out that cancer could be a genetic disease even if it were not inherited. Acquired as well as inborn genetic alterations play a major role in cancer, he noted, and in any case people differ in their inborn qualities in ways that are directly relevant to the control of cancer. "It is well-known, for example, that not all elderly men who smoked heavily most of their lives develop lung or other tobacco-related cancers. Why?" Some people, apparently, are born with genes that permit them to resist environmental mutation better than others (Fraumeni 2000).

This type of argument is powerful, ubiquitous, and applicable to many different bodily conditions. It provides an irrefutable defense: even if the environment is the primary cause of most cancers, the individual body's response to the environment is the crucial problem to be explained. For geneticists (understandably) the genome is the proper site at which cancer can be controlled.

As the *Newsweek* quotation that opens this chapter suggests, by 1984, in popular culture, all disease had become genetic disease. The technical and clinical realization of this perspective was still in process, but its ideological frame was in place. For the purposes of public discussion, then, all disease was genetic disease.

As I close my exploration of genetic disease in the postwar period, I want to suggest that all disease is not, in practice, any one thing. Disease is a multi-layered human experience, and its causes are diverse. It can be a product of water purification standards, bacteria, social disruption, poor health care, heredity, injustice, military action, viruses, mutations, traffic patterns, drought, emotional abuse, chemical imbalances, toxic waste, poison, malformation, smoking, childbirth, or diet. Two people can have the same disease for different reasons. Two people can have different diseases for the same reason. If a "cause" is something that if changed would prevent the phenomenon in question, genes emphatically do not cause all disease. It is true that some people are born with a protein anomaly that makes them unable to contract HIV. It is not true that the control of HIV should therefore focus on genetic enhancement. Even if people differ in their susceptibility to the virus, HIV/AIDS is not a genetic disease.

Why and how, then, has the idea that "all disease is genetic disease" come to play such a prominent role in popular and scientific culture? It was an idea realized and institutionalized long before technologies for mapping the human genome were available. The technologies were in some ways a consequence of the idea, rather than a cause, as new ways of manipulating DNA were valued because they would be relevant to the control of human disease and, not coincidentally, valuable in a medical marketplace increasingly driven by the prerogatives of the pharmaceutical and biotechnology industries. Genes "fit" the marketplace: they are small pieces of nature that can be patented, hoarded, distributed, mass-produced, and isolated. They conform physically to modern biomedicine, being perfectly suited to a particular institutional configuration of health care delivery, private industry, global commerce, and the contemporary academy. They are the right disease vectors for the early twenty-first century in the developed world.

As social historians of medicine have been demonstrating for the past forty years, illness is culturally and historically specific in revealing ways. This perspective in no way denies the biological reality of illness, the embodied truth of human suffering, or the efficacy and importance of medical intervention. It does, however, suggest that disease is a complicated biosocial experience, mediated by technologies of diagnosis and intervention and by causal and moral narratives that explain particular bodily states. The legitimacy of this perspective is amply suggested by the thick network of meanings built around HIV/AIDS, which is at one and the same time a biological, social, economic, political, and emotional phenomenon (Epstein

1996). But less serious illnesses such as Lyme disease also have complicated social meanings, as do states such as menopause or old age, which are diseases in practice, by consensus if not strictly by biology (Aronowitz 1998; Brumberg 1988). Explanations, technologies, interventions, perhaps even pathologies are different in different times and places. And in a single time and place, the same bodily condition can be subject to multiple and conflicting interpretations and meanings. Disease is often a place where a culture's moral narratives, social organization, and economic structures are made manifest. Just as a mutant fly is a window onto the genome, so disease is a window onto culture.

As my story suggests, the longings expressed around genomics are by no means confined to technical experts. Part of the reason that genomics has fared so well is that it intersects with older concerns about genealogy, relatedness, sameness, and causation. The rise of genetic disease to medical prominence has been a community project, reflecting broad social interest in explanations that could track the suffering caused by disease to some ultimate source.

In the thirty years from 1970 to 2000, the science of genetics was transformed. Scientific knowledge of heredity became a critical resource for the nascent biotechnology industry, and the industry in turn affected careers, professional standards, and scientific education. Technological capabilities also expanded. Genetic engineering, the manipulation of DNA to produce changes in the biology of living organisms, initially worried some scientists. One group called a special conference in 1975 to discuss how genetically engineered organisms should be handled and controlled. At the Asilomar Conference, scientists themselves sought to address the social and ethical implications of the new technological capabilities. Perhaps not surprisingly, they concluded that the risks could be controlled by consensus within the scientific community. This did not hold, however, and governments in Europe and elsewhere began to play an active role in the effort to control research in molecular genetics (Wright 1994; Gottweiss 1998).

In the 1980s, some geneticists began to think it might be possible to pursue what Victor McKusick had proposed in 1969: a complete map of the entire human genome. Genes in other organisms had been mapped as early as the 1910s by geneticists who carried out controlled crosses with the intention of locating genes along the chromosomes. But the mapping of the human genome being proposed in the early 1980s would be made possible by new technological capabilities that permitted direct study of the genetic material itself. Human DNA could, in theory, be sequenced until all of the estimated three billion nucleotides were known.

The U.S. Department of Energy began exploring the possibility of supporting such a project in 1986. The department, successor agency to the Atomic Energy Commission, had been an important source of support for biological research in the United States since the 1940s. The primary justification for this support related to the genetic effects of radiation, as the Atomic Energy Commission (later, Department of Energy) was charged with assessing the safety of the exposure of human populations to radiation in and around nuclear power plants and resulting from fallout and radioactive waste. One of the difficulties of tracking radiation damage had long been that the "natural" mutation rate in *Homo sapiens* was unknown. A map of the genome might shed light on the rate of normal or spontaneous mutation in human populations, which could then provide a baseline for the assessment of the added risk posed by radiation exposure. This relatively modest plan, to prepare a map as a guide to radiation risk, was rapidly replaced by a much more elaborate program with a broader medical agenda and with many other institutional participants (Cook-Deegan 1995).

In the mid 1980s, conferences on the possibility of constructing a map of the entire human genome were held at the NIH, the Cold Spring Harbor Laboratories, and the Imperial Cancer Research Fund Laboratories in London. In 1988 a group of geneticists founded the Human Genome Organization, an international professional group that coordinated work across laboratories and countries. They reached an agreement about how to divide up the mapping of the twenty-four human chromosomes (twenty-two matched pairs, and X and Y) to avoid duplication, and they sponsored a series of international workshops and meetings. The same year, the NIH created an Office of Genome Research and hired James Watson to run it. The genome began to be sold to Congress and the public as a fifteen-year endeavor that would have tremendous medical benefits and that deserved significant public funding.

One of the most heavily promoted future benefits of the genome project was gene therapy. Gene therapy could in theory permit the control of any genetic disease that could be mapped. There were two possible forms. Germline gene therapy would involve changing genes in eggs and sperm, such as replacing a defective gene with a functioning one. These changes would be-

MOMENTS OF TRUTH IN GENETIC MEDICINE

come a part of the human gene pool indefinitely, to be inherited by future generations with unknown consequences. Far more acceptable both in the scientific community and beyond was somatic gene therapy. This would entail inserting corrective genes into the cells of affected organs of people with genetic diseases. The cells, transformed by the new genes, could then begin to do what they were supposed to do; the genetic change would be confined to the recipient, not passed to offspring.

The FDA and other U.S. agencies began to explore such possible uses of recombinant DNA technology in the mid 1980s. Much of this discussion focused on how to control the risks in gene therapy research and how to control gene therapy researchers. This was partly a reaction to the activities of University of California, Los Angeles, hematologist Martin Cline in 1979–80. Cline bypassed the university's Institutional Review Board to carry out experimental gene therapy in Israel and Italy, trying to use the newly cloned globin gene to counteract thalassemia (www.georgetown.edu/research/nrebl/scopenotes/sn24.html). He was forced to resign his departmental chairmanship at the university and he forfeited some of his grants, eventually leaving the field of gene therapy.

In 1984 a new NIH committee was formed, the Human Gene Therapy Working Group, and ten years later, by 1994, the NIH had approved more than sixty applications for new gene therapy investigations. Proposals involved the use of viruses as vectors to deliver the working gene to the relevant cells. But abruptly, in the fall of 1999, gene therapy changed. The NIH ratcheted up its scrutiny, and the FDA suspended clinical trials at several institutions. This was the result of what happened to Jesse Gelsinger.

Jesse Gelsinger

In the fall of 1999, a young subject of gene therapy research died during an experiment at my home institution, the University of Pennsylvania. Jesse Gelsinger's death on 17 September 1999 had a broad impact on the community of researchers working in gene therapy. It led eventually to a lawsuit, derailed the career of the "gene therapy trailblazer" James Wilson, called into legal question the role of bioethicists in clinical trials, and transformed the Institute for Human Gene Therapy at Penn into a site for studies of animals rather than people. It provoked the NIH to revise its rules on safety in gene therapy trials. It led the FDA to halt all clinical trials at the institute at Penn and to assess more stringently clinical trials elsewhere.

Eventually, as the crisis unfolded in the fall of 1999 and the spring of 2000, it led to the revelation that many other gene therapy researchers were not reporting "adverse events" to the NIH and that other serious reactions to gene therapy had been kept quiet. There were allegations of at least six other deaths. Gelsinger's death challenged the technical and ethical infrastructure of genomic research. It also called global attention to the commercial networks through which so much genomic research moves: Wilson was founder and 30 percent owner of Genovo, a biotech firm that provided $2.8 million per year to the institute. The university itself was a shareholder as well and stood to gain royalties from any new gene therapies developed. The dean of the medical school held a patent on the adenovirus vector used.

When W. French Anderson of the NIH carried out his first clinical trial of human gene therapy in 1990, expectations for the technology were extremely high. Over the next decade researchers worldwide launched more than four hundred clinical trials of gene therapy. The creation of the "world's first institute for human gene therapy" was announced to the press in December 1992 at the University of Pennsylvania Medical Center. William N. Kelley, then chief executive officer of the Medical Center and dean of the School of Medicine, proclaimed that "there is increasing evidence that many of the most important breakthroughs in medicine will occur in the area of gene therapy, revolutionizing medicine as we know it today."[3] The institute's new director was to be "world-renowned geneticist" James M. Wilson, the first physician and scientist to conduct clinical trials of gene therapy outside the NIH. Wilson was the "Michael Jordan of gene therapy" and the "best person in the field."[4] In the press releases prepared by the university, Wilson's role in the biotech firm Genovo was not mentioned. His academic credentials and humanitarian goals were emphasized. His commercial stakes disappeared: the silences and omissions in the public narrative are as important as the things included.

As he settled in at Penn, Wilson began working on trials of gene therapy for familial hypercholesterolemia, ornithine transcarbamylase (OTC) deficiency, and cystic fibrosis. In April 1997 Wilson and his team began a phase I clinical trial of an experimental gene therapy for OTC deficiency. This was the condition Gelsinger had. It is an inherited disorder that leads to problems in processing the nitrogen in food proteins. Many affected males die in infancy. Gelsinger, who was eighteen when he died, had a relatively mild case and could manage the disease with diet and drugs. Like phase I trials

MOMENTS OF TRUTH IN GENETIC MEDICINE

more generally, the trial in which he died was a test of safety rather than of efficacy. He was the eighteenth patient to receive the adenovirus vector (www.uphs.upenn.edu/ihgt).

Later, Gelsinger's family felt that the informed consent materials used in the study and signed by eighteen-year-old Jesse were misleading. They expected a mild fever, and they thought that Jesse's participation would be crucial to the development of therapeutic interventions for others affected by OTC deficiency. They remembered being told that some treated patients actually had an increase in OTC activity, that the risks were remote, and that the therapy seemed to be working.[5] But four days after the adenovirus was injected, Gelsinger was dead of systemic organ failure that culminated in brain death. His immune system had reacted to the viral vector in devastating ways.

Even before Gelsinger died, gene therapy pioneer Anderson had grown openly skeptical about the potential of this technology. He proposed at a 1998 symposium that the scientific community had expected too much of gene therapy. Organisms, he said, "have spent thousands of years learning how to protect themselves from having exogenous DNA get into their genomes. So we were all a little naive to think that if we made a viral vector and put it into the human body it would work. The body's done a very good job of recognizing viral sequences and inactivating them."[6] In 1995, in the wake of some unimpressive early results with gene therapy, NIH director Harold Varmus appointed a committee to review and assess NIH's investments in the new technology. The resulting report stated that clinical efficacy had not yet been demonstrated, despite anecdotal claims of success, and "significant problems remain in all basic aspects of gene therapy. Major difficulties at the basic level include shortcomings in all current gene transfer vectors and an inadequate understanding of the biological interaction of these vectors with the host." In early 2000, Anderson told an FDA reporter that the "great burst of enthusiasm" in the early years had been tempered by the clinical results. "We came to realize that nothing was really working at the clinical level" (www.fda.gov/fdac/features/2000/500_gene .html).

Gene therapy, then, was a disappointment. After Gelsinger's death, both the FDA and the NIH investigated the work at the University of Pennsylvania and at other institutions where gene therapy trials were underway. News reports about the failure of scientists at several institutions to conform to

FDA protocols affected patients' cooperation: human subjects for trials of gene therapy suddenly lost interest. "Participation in gene therapy trials is way down, because the public is not sure what to make of this," said one FDA official (www.fda.gov/fdac/features/2000/500_gene.html).

The Gelsinger family sued the university, Wilson, and Penn bioethicist Arthur Caplan, among others, for wrongful death. The suit claimed that Caplan bore special responsibility because of his crucial role in encouraging the use of healthy adult volunteers rather than critically ill infants. Caplan had argued that informed consent was impossible for parents with a critically ill infant and urged the gene therapy group to work with healthy adults who had mild forms of the disease and from whom informed consent was a realistic option (www.sskrplaw.com/links/healthcare2.html). The case raised the question of legal liability in the growing bioethics industry. Could an ethicist be held accountable for problematic advice? The U.S. Senate held a hearing on gene therapy and the adequacy of public oversight of researchers (2 February 2000). The FDA increased the scrutiny of all gene therapy trials and began holding Gene Transfer Safety Symposia at which researchers involved in gene therapy could discuss their problems, learn all the rules, and learn from each other. Caplan's bioethics program, which had been a part of Wilson's department, was reconfigured as a separate, free-standing department in Penn's medical school. Wilson stopped working with human subjects and later stepped down as director of the institute, in the summer of 2002. He remained on the faculty at Penn; his career as a pioneering gene therapy researcher, however, was concluded. The lawsuit was settled, though the Gelsinger family expressed some dissatisfaction with Penn's responses.

In March 2000, only six months after Gelsinger's death, *Nature Genetics* published a very preliminary report of a modest clinical response in a gene therapy trial in patients with severe hemophilia. "Why is *Nature Genetics* publishing this study?" the issue's editorial asked. The answer was that despite the report's modest findings, it "may prove to be the first report of clinically efficacious application of gene therapy to hemophilia." The study was also published, the editorial said, because it was "exemplary in other respects." Researchers had used proper protocols in two appropriate animal models before applying the therapy to humans, they had carefully demonstrated an absence of vector-related toxicity, and they had compelling reasons to move into clinical trials. The "attractive study design" seemed to be as important as the resulting data in the decision to publish. *Nature Genetics* was in effect holding up the hemophilia study as an example of the way that human gene therapy research should be conducted, in the wake of the "heated debate since the tragic death of Jesse Gelsinger." All those working in gene therapy research should ask whether "any research protocol meets the litmus test of whether you would enroll yourself or your loved ones in this protocol given the same circumstance. This study meets that test."[7] Publication, then, was explicitly tied here to ethical research practices.

By 2002, some clinical trials of gene therapy, for hemophilia and severe combined immune deficiency, seemed promising, and gene therapy had by no means been abandoned—by the FDA, the NIH, or the network of academic and industry researchers interested in its potential. But many observers had become skeptical and disillusioned. Abby S. Meyers, president of the National Organization for Rare Disorders, an umbrella group of patients' organizations, expressed the general frustration. "We haven't even taken one baby step beyond that first clinical experiment. It has hardly gotten anywhere. Over the last ten years, I have been very disappointed" (quoted in Thompson 2000). The anticipation that gene therapy would provide a simple, clean solution for genetic disease proved incorrect.

James Wilson, promoting his gene therapy program in 1992, had contrasted the draconian surgical interventions pursued to treat one genetic disease, familial hypercholesterolemia, with the elegance and simplicity of gene therapy that could "use the patient's own cells."[8] In 1998, when a group of scientists and ethicists gathered at the University of California, Los Angeles, to discuss the potential of human germ-line genetic engineering, one of the key themes of the discussion was that such genetic engineering would be simple. Mario Capecchi of the University of Utah said germ-line therapy would be "much simpler" than somatic gene therapy; Leroy Hood of the University of Washington said that "in the long run we will essentially be doing . . . everything at the germline rather than in somatic tissues"; Lee Silver of Princeton University said that in germ-line therapy, "you can select one cell and produce the whole embryo from it, that's all you need to do."[9]

Yet simplicity, as my historical reconstruction has suggested, is itself a complicated cultural product. The idea that genetic disease somehow simplifies health problems is the end product of a process in which strategic silences and focused emphases have permitted some questions to become irrelevant and others to dominate public culture and technical work. It takes sustained effort to make anything as complicated as human disease appear simple, clear, or straightforward. Even the technical record of genomic medicine—the published scientific record—suggests that genetic disease is stunning in its complexity, even at the level of the cell. As one moves "up" to the organism, the environment, the species, and the social organization of medicine, research, family life, abortion, and political power, the complexity increases at every level. Part of the agenda of genomic medicine has been to promote the idea that clean, simple, technically sophisticated answers will come from genomics. A gene will be spliced into a defective genome to solve the problem once and for all. But many people have worked long and hard to build the complicated infrastructure that supports and sustains this idea of simplicity in public culture.

The People's Genome

The People's Genome Celebration in Washington, D.C., in June 2001 marked the completion of one phase of the mapping of the entire human genome. Organized by the Genetic Alliance, a network of genetic disease support groups, the celebration attracted people with genetic diseases, parents and family members, genetic counselors, physicians, genomics scientists, science policy professionals, science studies scholars like myself, and journalists. It featured presentations by the musician Todd Barton, who composed "Genome I," a symphony based on genetic patterns; by high-fashion photographer Rick Guidotti, who takes striking photos of people with albinism; and by the director of the National Human Genome Research Institute of the NIH, Francis Collins, who is not only a physician and geneticist but also a folk singer. Collins serenaded the group, playing guitar and singing a song he had written about genomics. As he repeated the performance several times over the course of the three-day event, many in attendance, including myself, were singing along: "This is a song, for all the good people, all the good people whose genome we celebrate . . . We're joined together by this common thread . . . We're grounded in science and the Genetic Alliance . . ." and so on. The People's Genome Celebration was festive, and the potential of genomics to transform medical care and alleviate human suffering was an important theme.

But a nurse from Connecticut who has children with sickle cell anemia articulated in her talk some of the tensions less openly acknowledged during the general festivities. Victoria Odesina noted that "we have known for over forty years about the gene for sickle cell anemia." The molecular mechanism, similarly, has been known for decades. Yet treatment for this painful and disabling condition is grueling and there is no cure. "Why not?" she asked. She agreed that, with completion of the first draft of the map of the human genome, it had been an exciting year for genomic science and for the community of support groups focused on genetic disease. But it was also important to notice that "we may not be able to find treatments or cures based on genetic information."[10]

Virtually all scientific and press reports about newly found genes include a sentence that proposes that finding a gene will lead to a cure for the relevant disease. This is ubiquitous enough to be understood as a literary convention in genomics. It is a statement or framing mechanism necessary to the structure of the argument. The proposal that the gene will lead to a cure is the primary explicit justification for the search for disease genes, and indeed for genomic technology more generally. But discovered genes do not lead directly to cures for people who have genetic diseases. Even journalists are starting to notice. Even Collins has become publicly, openly concerned about the need for a real clinical payoff from gene mapping; he has appointed a medical geneticist, Alan Guttmacher, to focus specifically on developing clinical applications of the knowledge produced by the Human Genome Project (Taussig 2005). And even parents driven to hunt for genes have acknowledged that finding the gene does not cure their children. Brad Margus, the shrimp merchant turned biotech chief executive who has two sons with the fatal genetic disease ataxia telangiectasia, helped raise money to find the gene, which was successfully identified in 1995. But in 2000 he told a reporter for the *San Jose Mercury News,* "Every time I see them [his sons], I know we haven't done anything, because we haven't stopped the progression. My kids are slipping away."[11]

In practice, discovered genes might shed light on the causes and processes involved in a complex genetic disease or in a disease with some genetic component. Although such information is useful, it is not easily translated into effective therapeutic interventions. In most cases, discovered genes have the immediate potential only to become diagnostic genetic tests. If a condition is devastating enough, and if there is a known population at risk to whom the test can be marketed, a discovered gene can be applied to fetal testing within a few weeks. Genetic tests can be sold to at-risk consumers as they make reproductive decisions, before marriage, before pregnancy, before implantation, before the twelfth week of gestation. Genomics is an early warning system: genes are things that will happen, that can be seen before they happen. There are not many things like that.

Abortion—the "A" word, as one prominent genomics scientist called it in a public lecture in 2002—is the covert technology of the Human Genome Project. Sequencers, polymerase chain reaction, fluorescent tagging, and DNA fingerprinting are ascribed deep significance in press and popular accounts of genomics and appear as unproblematic methods for technical elucidation in published scientific papers. Abortion, the actual technological means by which genetic disease is coming under greater control, is rarely mentioned. Indeed, it is rarely mentioned even by outspoken critics of genomic technologies, who call attention to the ethical issues of privacy, access to health care insurance, the risks of cloning and genetic engineering, and the "new eugenics" of genetic science, but who rarely attend to the discordance between the ways in which genomic medicine is promoted (as a guide to cure) and the ways it is practiced.

The first line of defense in the control of genetic disease, however, remains the selective abortion of affected fetuses. Given the complexity of the genome and the body, and the simplicity of abortion, this dynamic can be expected to continue.

When I was preparing a short essay on this project for the *American Journal of Medical Genetics* in 2001, I included a sentence stating that the selective abortion of affected fetuses was the primary intervention of genomic medicine. In an email, an editor at the journal asked me to change the wording, to say instead that the selective abortion of affected fetuses was "a common intervention." I did not respond to the email message. I decided that if the editors chose to change the wording, I would not object. But neither did I want to explicitly approve the change. I ceded the decision to them, and, perhaps because of my silence, they left the sentence as I had written it. It appeared that way in the published paper (Lindee 2002).

My silence in response to that email message was shaped by the complexity of the issues and perhaps even by my own conflicting reactions to abortion and the abortion debate in the United States. I considered the rewording to be somewhat misleading, but I understood why the editors would be concerned. I am a feminist supporter of abortion rights, and I know enough about genetic disease to recognize the terrible impact that the birth of a child with a severe or fatal disease can have on a family, particu-

MOMENTS OF TRUTH IN GENETIC MEDICINE

larly in a culture with limited tolerance for disability, an ideology of human perfectibility, less-than-universal access to medical care, and limited support for families that must care for seriously ill or disabled children. At the same time, I am perplexed and troubled by the disconnect between the public promotion of genomic medicine, in which research is supposed to lead to cures, and the practical realities of limited clinical options. The allocation of significant public and private resources to genomics has proceeded within a network of public silences, perhaps widely accepted. Maybe an early warning system is what we want, not only for reproduction but for self-maintenance. Maybe we want to know that we are at risk for cancer so that we can improve our diets or consume whatever preventive pharmaceuticals might be made available by the biotechnology industry. Maybe we want quality control for the next generation. Certainly it is what we are getting, and I think it is important to say so in plain language.

As diagnostic capabilities have grown increasingly baroque, fetuses can, in theory, be screened for hundreds of genes. Some of these genes are associated with conditions that are highly variable, have late onset, or are heavily mediated by behavior and environment. By 1998, for example, there were seventeen candidate genes for obesity in the published literature (Comuzzi and Allison 1998). If prospective parents had the knowledge, if they had the option, would they choose to bear a child who had all seventeen? Given the moral and social meanings of weight, the connections between bodily morphology and professional success, and the pressures on parents to produce perfect children, how far away is selection against "fat genes"?

In my introduction in chapter 1, I proposed that people presumably differ in the rates at which they starve to death but that this fact, however true, would not make starvation a genetic disease. Everyone, eventually, will starve to death if deprived of food. Genes do not "cause" starvation. Nor can genes prevent starvation. Starvation is a consequence of food deprivation. The same logic applies to the morphological opposite of starvation, obesity. People unquestionably differ in the rates at which they become obese, and these differences presumably reflect, at some level, genetic variations in metabolism or some other quality. But eventually, everyone will become obese if they consume enough food. Obesity researchers know this. Everyone has the potential to become obese with sufficient caloric intake. Genes can neither cause (in any simple sense) nor prevent obesity. Obesity is a consequence of unlimited access to food.

I was interested in pursuing this comparison, so I searched the online database of medical and scientific publications, Medline, for the term *starvation gene*. I found seven citations, six focusing on the effects of an inadequate nutrient medium on gene expression in bacteria, one on a mutant *Escherichia coli* strain with a "starvation gene" that compromised survival under conditions of deprived nutrition. None of these studies focused on human starvation genes. When I tried *obesity genes*, I got more than 2.6 million hits. This seemed excessive, so I modified the search with *human*, which reduced the list to 514,876 citations. Obesity is a disease of prosperous nations that affects millions of consumers who can afford to pay for drug interventions. Starvation affects millions of people who have limited resources. What kinds of questions loom large in genomic medicine? What conditions come to be understood in genetic terms, and why?

Certainly underlying the appeal of the obesity gene is the idea that a trait that is genetic is somehow easier to manage than something that requires changes in the environment or in individual behavior. It is difficult to control food intake under conditions of unlimited access, as any dieter knows. Also underlying the appeal is the marketplace value of weight loss: an effective pharmaceutical intervention for obesity would be extremely profitable. And genes are widgets, pieces of nature that fit neatly into a market economy.

Biotechnology in the United States grew from a $5 billion industry in 1989 to a $25 billion industry in 2000. The number of patents pending for human DNA sequences has grown from four thousand in 1991 to five hundred thousand in 1998 to several million in 2004. Genomic medicine is corporate medicine, driven by the prerogatives of the biotech industry.

In this study, I have explored the people, institutions, and ideas that facilitated the reconfiguration of human disease in genetic terms in the United States in the long 1960s. I focused particularly on a relatively short but critical period, from about 1955 until about 1975, during which genetic disease was transformed in many ways. I have suggested that something remarkable happened around 1959 or 1960, when a whole range of intellectual and political shifts brought genetic disease to medical and public attention in dramatic ways. Always I have asked how and why all disease came to be understood as genetic disease, at least for the purposes of public discussion. I have tried to stay close to the tangible manifestations of this idea, to watch how it unfolded in institutional and clinical settings. I have tried to notice who was doing the work of making it a reality.

What can my story suggest? First, knowledge is socially produced at many levels. What is acceptable or thinkable at a given time becomes a resource for technical knowledge production, and trust relationships extend much more broadly than might seem obvious. People can and do report reliable knowledge based on their embodied social experience. Many different moments of truth are bound up in technical reports. At the same time, there may be things that cannot be technically known because they are not known or experienced in the context of a given culture. There may be ways of seeing the body and experiencing disease that are absent not only from social life but from technical interpretations, because they are not experienced culturally in industrialized societies.

Genomics may be a technical elaboration of a well-established, much older way of seeing bodies, risks, and relationships, grounded in an industrialized model of efficient management. Technical knowledge and folk and emotional knowledge all are engaged with genetic disease, and the rise of genetic disease to medical prominence has been a complicated group project. The dreams fueling genomics have come from patients and parents, mental health professionals and technicians, administrators and legislators. The data themselves have often been seen to be true by people credentialed by bodily experience: possessed genes and possessed knowledge may overlap. Embodiment teaches reliable things, and people know things in ways that are not scientific but can be taken up in scientific narratives to explain natural phenomena. People choose to interpret bodily experiences in ways that reflect prevailing cultural norms, and their choices are critical to the enterprise.

Genomic medicine also reflects, at every level, the effects of corporate culture on biology in the second half of the twentieth century. It expresses the expectation that the body can and should be subject to very high levels of surveillance—fetal testing, preemptive testing for BRCA genes (associated with breast cancer), or testing for Huntington disease or other genetic disorders. The body needs to be tracked, inspected, and constantly monitored in the hopes of preserving health. An early warning system in which predictive information can be put to use appears to be inherently valuable: there is a widespread presumption that any rational person would prefer to know such information, and such information should be used consistently to make reproductive and health decisions. The idea that all disease is genetic disease encodes a model of embodied experience in which technological surveillance is morally and socially desirable and necessary. It also draws on the conviction that technological interventions are now or soon will be able to change or avoid certain futures. The outcomes predicted by the early warning system, in this model, will become plastic, fungible, changeable in response to individual desire. This emerging model of genetic disease promises the substitution of one future event for another. This is not a passive way of thinking about bodies or health. It is an aggressive form of technological optimism.

In his talk at the People's Genome Celebration, Francis Collins sketched out his vision of the next thirty years of genomic medicine. By 2010, he proposed, predictive tests would be widely available for a dozen or so complicated conditions, and interventions would be available, too. Gene therapy would be used effectively, and primary care physicians would be practicing genetic medicine. There would also be effective federal legislation to solve problems of discrimination, privacy, and ethics, but access to genetic services would remain inequitable. By 2020, Collins said, more genomic drugs would be on the market and cancer would be controlled by a targeted drug keyed to the molecular fingerprint of the relevant tumor. At least 120 cancer drugs premised on this possibility were already in the pipeline, he said. And mental illness, by 2020, will have "many of its secrets revealed." By 2030, genomic medicine will be the norm. Individual preventive medicine based on genomic data will guide human behavior (diet, exercise, drugs), and the environment-genotype interaction will be understood, so that information from the genome can be applied directly to clinical care. The average life span by 2030 will be ninety years, though all this progress will be threatened by a major antitechnology movement that will be active worldwide. As Collins presented it, a transformation in health care delivery, in which health care comes to be understood as a right, not a privilege, would provide a compelling antidote to the "irrational concerns of the anti-technology groups."[12]

Environmentalists have indeed already begun to turn their attention to genetic technologies, partly as a result of the global debate on human cloning, and their concerns do not sound irrational. In a summer 2002 issue of *World Watch* devoted to the question of cloning, Richard Hayes, of

MOMENTS OF TRUTH IN GENETIC MEDICINE

the Center for Genetics and Society in Oakland, California, proposed that the "new human genetic technologies are arguably the most consequential technologies ever developed." These technologies could prevent disease and alleviate suffering, and also destabilize human biology and undermine the foundations of civil society. "If cloning is the atomic bomb of the new human genetic technologies," he proposed, "IGM [inheritable genetic modification, or germ-line gene therapy] is the multi-megaton hydrogen bomb. Only the most egotistical or deluded would want to clone themselves, but if IGM were allowed even many who are appalled at the prospect of using it would feel compelled to do so, lest their children be left behind in the new techno-eugenic rat-race" (Hayes 2002, 12). Writing in the same issue of *World Watch*, Sarah Sexton of the Corner House, a U.K.-based research group focused on social and environmental justice issues, observed that "the great majority of the world's diseases are caused by environmental, not genetic, conditions. A frenzied search for genetic therapies could steal resources from billions in order to serve only a few" (Sexton 2002, 18). Spurred by the growing fascination with genes, she proposed, policymakers and the public have come to see medicine primarily as a process of "fixing" diseased individuals and good health as "something to be bought and sold in the marketplace by individual consumers, rather than as a political goal for society to work toward" (19).

The Wellcome Trust, which has a significant stake in genomic science, recently funded a workshop to explore how patients and research subjects can be integrated into scientific policy setting and decision making. This was in response to a British House of Lords proposal in February 2000 that the public should play an integral role in science policy making. Consumers, at least in Britain, are "more questioning" and "regard expert input as advice or opinion to be weighed up against other points of view." Disease-specific charities have been in the vanguard in their promotion of consumer involvement. The British Alzheimer's Society, for example, enrolls consumers to review grant applications, sit on award panels, and set strategies for funding. "Researchers say that the most penetrating questions often come from the consumer members of the award panel." Such a model involves not "public understanding of science" but "public ownership of science."[13] But consumers have perhaps played an important role in biomedical science for decades.

In 2000, the first genetically modified animal intended for human consumption, a rapidly growing salmon produced by the Boston-based A/F Protein Inc., was submitted for FDA approval for marketing in the United States. Nicknamed Frankensalmon, the fish was the focus of a sustained critique by environmentalists. Also in the works is a genetically modified pig that will produce leaner meat. Indeed, the genetic modification of farm animals has proceeded rapidly over the past decade. Chickens, cows, goats, sheep, and pigs have been the agricultural models, the first wave, in the genetic modification revolution.

At the turn of the last century, from 1880 to 1920, such farm animals provided models for human eugenics. The eugenics literature was filled with allusions to "racehorses and drafthorses" and to the evidentiary power of the well-bred cow. The benefits of improved breeding for human populations were extrapolated from the benefits manifest in dogs, sheep, cattle, and other domesticated organisms. These animals were the stand-ins, the ones who experienced eugenics first. As the subjects of controlled breeding by farmers and breeders, they demonstrated the potential of planned reproductive control. So, too, as the genomics revolution proceeds, genetically modified animals and plants provide a model, a sign, a first take on the potential transformation of human beings. Their fate provides an image of a possible human future. The cloned sheep Dolly is a critical artifact, and her behavior, reproduction, and eventual death constitute a rehearsal and a trans-species warning signal (Franklin 2003). The first cloned kitten, CC, was marketed as a domestic pet, an emotional technology, and a surrogate child brought into being through cloning (www.savingsandclone.com).

The "old eugenics" was left behind in the industrialized world in fits and starts, sometime between about 1950 and 1980. It had devastating effects for individuals, it was used to justify genocide, and in some forms constituted a serious breach of human rights. In Nazi Germany it was implicated in mass murder. But it was inefficient, in terms of the gene pool. It simply did not work. The cumulative impact of the old eugenics on the traits circulating in the human gene pool was almost certainly negligible. The species *Homo sapiens* in 1950 was the same species it had been in 1900, neither better nor worse. Eugenics did not eliminate mental retardation or criminality. It did not lead to a future of phenotypically attractive and fit families, tall, well-boned, with an industrious work ethic. Things ended up, biologically, much as they had been.

The new eugenics, promising germ-line gene therapy and even genetic improvement of the human species, is much more potent biologically. It could actually work. It could affect the gene pool. It is true that gene therapy has not yet worked very well and cloning is inefficient and dangerous. But if these technologies are ever made to work in human reproduction, the consequences could be expected to play out over a long, long time.

The unpredictability of technical intervention is a powerful theme in the history of science, technology, and medicine. Even the most beneficial interventions—and here antibiotics would be a signal case study—have often had unexpected adverse consequences. The widespread use of antibiotics, particularly in animal feed, has produced resistant strains of pathogens. This is not an argument against the development and use of antibiotics. It does suggest that antibiotics could have been treated more carefully, their beneficial qualities "protected" with an awareness of the risks of resistance. Other technological interventions have had more devastating consequences and fewer benefits.

Germ-line genetic intervention might be like antibiotics: potentially transformative, offering new options for therapy that could improve many lives, but also having the potential to result in long-term damage. But it could also be like DDT, which won a Nobel Prize for Paul Müller, killed insects extremely well, and caused profound environmental damage. The potential long-term damage from germ-line therapy is opaque. We do not know exactly what could go wrong, and our hypotheses may well prove incorrect. Such long-term damage is also extremely likely, given the historical patterns of dramatic technological innovation: most innovations that bring benefits also prove costly, at some level, in some way. This is almost a "law of history." Germ-line gene therapy could introduce into the gene pool tremendous suffering, even if it were well-conceived and executed. There need not have been a technical mistake. The mistake, perhaps, would be in the social and political logic that permitted such an intervention to proceed.

We are therefore poised in the midst of another moment of truth, in which consumers and observers and producers of genomic knowledge are coming to terms with the direction of genomic medicine. This "moment of truth" is the confrontation between the capacities of the new technology and the embodied experiences of parenting and health care. In only a few generations, reproductive options could be drastically different. Our children, grandchildren, and great-grandchildren might be offered choices that would seem now frightening in their moral complexity and their long-term implications for society.

Genomic medicine, however, is consumer science and domestic science, and consumers will play a critical role in its development. The hesitation—the moment of doubt, even the moment of truth—could be the initial point at which it veers onto another course.

NOTES

ABBREVIATIONS

APS   American Philosophical Society, Philadelphia.
EVC 8005   "Ellis–van Creveld 8005," box R115 F1–2, Papers of
                        Victor A. McKusick, Alan Mason Chesney Medical
                        Archives, Johns Hopkins University, Baltimore.
JVN   Papers of James V. Neel, American Philosophical
                        Society, Philadelphia.
Medical Sciences Com.   Medical Sciences Committee on Inborn Metabolic
                        Disorders Archives of the National Academy of
                        Sciences, Washington D.C.
NARA I   National Archives and Records Administration,
                        Washington, D.C.
NARA II   National Archives and Records Administration,
                        College Park, Md.
NAS   Archives of the National Academy of Sciences,
                        Washington D.C.
USCB   Records of the U.S. Children's Bureau, National
                        Archives and Records Administration, College Park,
                        Md.
VAM   Papers of Victor A. McKusick, Alan Mason Chesney
                        Medical Archives, Johns Hopkins University,
                        Baltimore.

CHAPTER 1. INTRODUCTION

1. Twins seem to be right about their zygosity 90 percent of the time; DNA fingerprinting would produce more accurate results, but like other technical measures it would be too costly for large samples of unselected twins.

2. There is some relevant discussion of the history of ideas about like begets like in Medin and Atran 1999. See also the remarkable interpretation of Olmec knowledge by Tate and Bendersky (1999).

3. James Secord's 1981 paper remains a fresh exploration of Darwin's deep engagement with amateur breeders and the role of their knowledge of artificial selection in his thinking.

4. On the role of knowledge in agriculture and the contentious question of farmers' knowledge, see Hamlin and Shepard 1993.

5. J. Langdon Down's 1866 article, "Observations on an Ethnic Classification of Idiots," was reprinted in 1995 in *Mental Retardation*. See also Zihni 1995. On Huntington disease, Alice Wexler points out that physicians described the symptoms of excessive involuntary movement in earlier texts, and the seventeenth-century physician Thomas Sydenham tried to classify choreic syndromes, though he did not see them as genetic. In the 1840s, physicians in the United States, Norway, and England began to write about what they called *chronic hereditary chorea* (Wexler 1995, 46–49).

6. Weiner apparently ran his own lab, Weiner Serum Laboratory, and also had an appointment at the New York University Medical School. Later he was a bacteriologist at the Office of the Chief Medical Examiner of New York City. He was also involved during the war with the Office of Scientific Research and Development, the organization that mobilized American scientists and engineers to help the Allies during World War II (Cattell 1955).

7. I am estimating the number of amniocentesis procedures based on a 1967 report by Harold P. Klinger and Orlando J. Miller at a conference in Puerto Rico. They reported four successful diagnoses and five additional cases that had recently been reported (Klinger and Miller 1968).

8. From the comments of Dr. Richard Heller, director of the Prenatal Birth Defects Center, before the Maryland Commission on Hereditary Disorders meeting, 26 June 1974, minutes filed in Department of Health, Baltimore.

9. Ibid.

10. Collins made this comment at the People's Genome Celebration in Washington, D.C., June 2001.

11. The topic of limitations of clinical interventions came up several times in different contexts at a meeting at Harvard Medical School, 23–26 June 2001. The meeting was supported by the Applera Foundation, which was created by Applied Biosystems and its subsidiary company Celera, Craig Venter's gene mapping group. Leon Eisenberg organized the meeting, based on the premise that the average physician-

in-training needed to know much more about genomic medicine. One of the proposals was to abandon the teaching of anatomy in favor of more training in the anatomy of the genome.

12. For an excellent, comprehensive, and continually updated web resource on developments in genomic medicine, see www.genetics-and-society.org. Richard Hayes, Marcy Darnovsky, and their staff provide a critical public service through this very informative site.

13. Sulston's comments made at the People's Genome Celebration, June 2001.

CHAPTER 2. BABIES' BLOOD

1. For a description of the field trials, see Children's Bureau Publication No. 419, *Phenylketonuria: Detection in the Newborn Infant as a Routine Hospital Procedure*, U.S. Department of Health, Education and Welfare, 1965, copy on file in RG 287, box FS750, Publications of the U.S. Government, National Archives and Records Administration, Washington, D.C. (hereafter, NARA I). On the status of neonatal testing in late 1965, see Rudolph P. Hormuth, U.S. Children's Bureau, to M. Beaudet, 6 December 1965, 20-212 1963–66, box 1097, NN3-102, records of the U.S. Children's Bureau (hereafter, USCB), Central File, RG 102, National Archives and Records Administration, College Park, Md. (hereafter, NARA II).

2. Guthrie, "Phenylketonuria Detection in the Young Infant as a Routine Hospital Procedure," 2 May 1962, in 20-212, "Study of Phenylketonuria," box 950, USCB, Central File, RG 102, NARA II.

3. "Obtaining Blood Specimen for Phenylalanine Test (Guthrie Test)," in memorandum, William J. Peebles to Local Health Officers, 13 March 1968, Maryland Department of Health, Baltimore.

4. Maryland Department of Health, "Progress Report No. 8: Guthrie Screening Program for PKU," *PKU Newsletter*, 3 November 1965, records of the Maryland Department of Health, Baltimore.

5. Comments of Gladys Krueger at the Division of Health Services Staff Meeting, 6 June 1963, typescript in 20-212, 1963–66, box 1097, USCB, Central File, RG 102, NARA II.

6. Robert Guthrie, "Phenylketonuria Detection in the Young Infant as a Routine Hospital Procedure; A Trial of a Phenylalanine Screening Method in 400,000 Infants; a Proposal Submitted to the U.S. Children's Bureau 2 May 1962"; reprinted in full as Exhibit II in Children's Bureau Publication No. 419, *Phenylketonuria*, NARA I.

7. Other points included that the state health department should advise Guthrie of the number of babies they expected to screen, and Guthrie would then send the required number of test kits to state health departments. Also, the two-week urine follow-up was crucial both to catch missed cases and to assess the Guthrie test: "The two-week follow up urine phenylalanine (not phenyketone) test has not been em-

phasized but is of critical important [sic]. This test should (a) find *all* cases of PKU and (b) serve to evaluate the efficiency of the first blood phenylalanine test done on specimens take [sic] as late as possible when the infant is in the hospital nursery." Region II office of the Children's Bureau, John M. Saunders, M.D., Acting Regional Medical Director, CB, New York Office, 31 May 1962, in 20-212, "Study of Phenylke-tonuria," USCB, box 950, RG 102, NARA II.

8. Marsh to Hormuth, Specialist in Services for Mentally Retarded Children at the U.S. Children's Bureau, memo, re "Plan Material for Trial of Phenylalanine Screen Method in 400,000 Infants," 1 June 1962, in 20-212, "Study of Phenylke-tonuria," USCB, box 950, RG 102, NARA II.

9. Hormuth to Marsh, 6 June 1962, in 20-212, "Study of Phenylketonuria," USCB, box 950, RG 102, NARA II.

10. Rudolph P. Hormuth, Specialist in Services for Mentally Retarded Children at the U.S. Children's Bureau, to Ruby G. Martin, Staff Attorney, U.S. Civil Rights Commission, 11 April 1963, re "Exclusion of Non-white Children from PKU Screen-ing Program (Springfield, Massachusetts)," in 20-212 1963–66, box 1097, NN3-102, USCB, Central File, RG 102, NARA II.

11. In practice, in participating hospitals, Massachusetts was apparently screening all infants, white and nonwhite, and was soon to approve legislation requiring that all newborns be screened in all hospitals. The case that Martin referred to in Springfield, Massachusetts, was probably, said Hormuth, unrelated to the U.S. Children's Bureau program as it did not involve a newborn. "I gather from what you said that the situa-tion involved a mother who knew her child was mentally retarded and wished an eval-uation of whether or not the retardation was due to PKU." Screening of older children required a simple urine test that could be carried out by any physician or well-baby clinic, Hormuth said. "It may well be that this mother attempted to obtain this kind of screening from one of the hospitals which was participating in the newborn infant screening (requiring blood sample) rather than her own physician or health depart-ment facility for this screening test. The mother may have misunderstood what she was told about the purpose of the field trial program involving newborn infants" (ibid.).

12. Katherine B. Oettinger, Chief, Children's Bureau, to Herschel Cleaner, Inven-tions Coordinator at the Office of the Surgeon General, U.S. Public Health Service, 5 November 1963, re "Miles Laboratory Request for Exclusive Commercial Arrange-ment to Develop Guthrie PKU Test," in 20-212 1963–66, box 1097, NN3-102, USCB, Central File, RG 102, NARA II.

13. "Phenylketonuria," *Currents in Public Health* 2, no. 2 (1962). The "value to the child and to his parents is, of course, immeasurable." See also Centerwall, Chinnock, and Pusavat 1960. They estimated $20 million total savings over the lifetimes of two hundred infants with detected PKU. Their research, incidentally, was supported in part by Mead Johnson, which made one version of the low-phenylalanine formula.

14. Copy of brochures from Fergus Falls, Minnesota, and Findlay, Ohio, undated, with letter, Hormuth to William Swatek, 2 October 1962, in 20-212, "Study of Phenylketonuria," USCB, box 950, RG 102, NARA II.

15. "Procedures Used in Massachusetts PKU Screening Program," reprinted in Children's Bureau Publication No. 419, *Phenylketonuria*, 64–73.

16. Krueger, 1 February 1962, "Preliminary Planning of Cooperative Follow-up Study of Children with PKU," in 20-212, "Study of Phenylketonuria," USCB, box 950, RG 102, NARA II.

17. Ibid. Federal maternal and child health and "crippled children's" funds had been used to help support treatment programs for children with PKU in these clinics, and since 1959 these funds had been made available to help pay for the diet when the parents and clinics were unable to provide for it.

18. Ibid.

19. Letter from Arthur J. Lesser, Director, Division of Health Services, in 20-212, "Study of Phenylketonuria," USCB, box 950, RG 102, NARA II.

20. The form, dated 30 November 1962, "Preliminary Draft Form I, Clinical Study of Children with Phenylketonuria," was meant to be filled out when the children initially came to the attention of the clinic; in 20-212, "Study of Phenylketonuria," USCB, box 950, RG 102, NARA II.

21. "Preliminary Draft Form II, Clinical Study of Children with Phenylketonuria," in 20-212, "Study of Phenylketonuria," USCB, box 950, RG 102, NARA II.

22. Gladys M. Krueger to Marian M. Crane, 25 January 1963, in 20-212 1963–66, box 1097, NN3-102, USCB, Central File, RG 102, NARA II.

23. Ibid.

24. 20-212, "Study of Phenylketonuria," USCB, box 950, RG 102, NARA II.

25. Medical and Chirurgical Faculty of the State of Maryland, "Report of the Ad Hoc Committee for the Study of PKU Legislation," 4 April 1968, records of the Advisory Committee on Metabolic and Hereditary Disease, Department of Health, State of Maryland, Baltimore.

26. Benjamin White, chief of the division of community services for the mentally retarded of the Maryland Department of Health, to Madeleine Morcy, regional medical director at the Children's Bureau office in Charlottesville, Virginia, in unused cabinet near the offices of Marion Robertson, Department of Health, State of Maryland, Baltimore.

27. Benjamin White to Madeleine E. Morcy, 31 January 1963, "Division of Hereditary Disorders: PKU Correspondence and Legislation, 1963–1968," Maryland.

28. White to Dr. M. McKendree Boyer, and White to Dr. William M. Stifler, both 8 November 1963, "Division of Hereditary Disorders," Maryland.

29. Roney to Perry Prather, 27 May 1964, "Division of Hereditary Disorders," Maryland.

30. Prather to Roney, 5 June 1964, "Division of Hereditary Disorders," Maryland.

31. Edward Davens to James S. McAuliffe, Jr., 31 August 1964, "Division of Hereditary Disorders," Maryland.

32. The group drafted a resolution that was sent out to physicians' groups around the state. The resolution asked each group to formally endorse the Department of Health's "assumption of responsibility" for carrying out PKU testing. The department agreed to introduce a program of PKU screening into all Maryland hospitals, to work to develop other screening techniques for conditions other than PKU, to inform the public about PKU and other metabolic disorders, and to keep a register of all PKU patients. This was a resolution, in other words, that vested full responsibility for the program in the Maryland Department of Health, without mandating anything. "Resolution on State-Wide Screening for PKU and Other Conditions Leading to Mental Retardation," undated, "Division of Hereditary Disorders," Maryland.

33. Benjamin White to Edmund Bradley, 9 December 1964, "Division of Hereditary Disorders," Maryland.

34. Untitled minutes, "Division of Hereditary Disorders," Maryland.

35. Robert E. Cooke, M.D., Johns Hopkins University, Given Foundation Professor of Pediatrics, 28 December 1964, to Benjamin White, Chief, Division of Community Services for the Mentally Retarded, State Department of Health, Baltimore, "Division of Hereditary Disorders," Maryland.

36. Untitled minutes, "Division of Hereditary Disorders," Maryland.

37. Maryland Department of Health, *PKU Newsletter*, 29 December 1965, copies in "Division of Hereditary Disorders," Maryland.

38. Robert Guthrie, Children's Bureau Publication No. 419, copy of a memorandum of 9 January 1964 to all maternal and child health directors and laboratory directors participating in the trial of the inhibition assay for early detection of PKU, "Detection of Maternal Phenylketonuria," 60, 61.

39. Social Issues Committee, American Society of Human Genetics, "Large Scale Screening for Metabolic Defects," in correspondence, Arno Motulsky to Philip Handler, 6 March 1970, Medical Sciences Com. on Inborn Metabolic Disorders (hereafter, Medical Sciences Com.), Proposed, 1970, Archives of the National Academy of Sciences, Washington D.C. (hereafter, Archives NAS).

40. Motulsky to Handler, 6 March 1970, Medical Sciences Com., Proposed, 1970, Archives NAS.

41. Charles Dunham to the files, "Detection and Correction of Inborn Metabolic Defects," 4 December 1970, Medical Sciences Com., Proposed, 1970, Archives NAS.

42. The proposal itself ("Proposal for the Study of Inborn Errors of Metabolism," 13 December 1971) and accompanying correspondence are in Medical Sciences Com., 1971, Archives NAS. The proposal was submitted by Charles L. Dunham, chair of the Division of Medical Sciences of the NAS, and it requested $90,000 per year for two years.

43. Ibid.

44. Ibid. This figure of 6 percent has an intriguing later history. It appears in texts in the late 1980s as the percentage of all hospitalizations that could be attributed to straightforward genetic disease, though clearly that is not what it refers to here. In any case, this is an important iteration of the idea that genetic disease proper is implicated in a far larger proportion of hospitalization costs than most observers recognized. The idea of the "hidden cost" of genetic disease is folded into this construction. Although the eugenics literature earlier in the century routinely delineated hidden costs of hereditary weakness, I think this later construction, emphasizing specific diseases and focusing on the economics of hospitalization, presents a different sort of argument, one focused on the implications of genetic disease for medical practice. It is an implicit argument in favor of more training in human genetics for physicians, a program widely supported by prominent members of the American Society of Human Genetics.

45. PKU was the most important focus of the committee's work. The bulk of the text and appendices dealt with PKU testing. Other topics explored included questions of mandatory versus voluntary screening and a survey of health professionals.

46. http://genes-r-us.uthscsa.edu/resources/newborn/screenstatus.htm (accessed 7 June 2004).

CHAPTER 3. PROVENANCE AND THE PEDIGREE

1. Oxford English Dictionary, online. Thus, as stated in the *Empire Forestry Journal* 12 (1933): 198, the "problem of seed origin embraces both the geographical location where the seed was collected . . . and the genetic character of the mother trees . . . In European literature the term 'provenance' . . . has come to be used for the first phase of the problem."

2. A Medline search for publications for the years 1966–2002 turned up 268 papers on the Amish specifically and 134,493 with references to Amish genetic disease. Many of these papers focused on bipolar disorder.

3. Before 1924, Davenport organized dozens of field studies of albinos in Massachusetts, juvenile delinquents in Chicago, even of the Amish in Pennsylvania, but field methods were relatively casual (Kevles 1985, 55, 199–200).

4. Materials in box R115F1–2, Papers of Victor A. McKusick, Alan Mason Chesney Medical Archives, Johns Hopkins University, Baltimore (hereafter, VAM).

5. As David Brown has put it, McKusick "moved genetics out of the world of yeast and mice and into the world of clinical medicine." McKusick was not the only person involved in this movement, but he was extremely important to the field in the United States. See Brown, "A Profile of Victor McKusick," unpublished manuscript, VAM.

6. Ibid.

7. The first publication listed by McKusick in his bibliography of his publications

(in VAM) is "Broedel's Ulnar Palsy, with Unpublished Broedel Sketches," *Bulletin of the History of Medicine* 23 (1949): 469–79.

8. McKusick, bibliography, VAM.

9. "A Genetic, Anthropologic and Clinical Study of a Geographic Isolate," proposal, undated but written between April 1957 and June 1958, probably in late 1957, box R109C8, VAM.

10. Ibid., 3.

11. Most of the biochemical testing was contracted out. McKusick's team, consisting of himself and another physician, six medical students, two secretaries, two general assistants, and various consultants, conducted the laborious examinations and collection of materials, but the blood and cholesterol tests were sent out (ibid., 9).

12. Brown, "Profile of Victor McKusick," 33. Brown's profile is a fairly detailed biographical study of McKusick's life and work. McKusick tells the same story in the introduction to his edited book of 1978. I also ran across it in a later interview with him in a Hopkins publication.

13. McKusick, "Medical Genetic Studies of the Amish," grant application to National Foundation–March of Dimes, for 1978–80, "National Foundation 1979–80," box R109C16, VAM.

14. Ibid.

15. McKusick to Hostetler, 7 May 1963, "The Amish Population," box R115F1–2, VAM.

16. McKusick discusses these advantages of the Amish in many sources, but a specific list of "characteristics of Amish society Favorable for Genetic Studies" is in "Ellis–van Creveld 8005," box R115 F1–2, VAM (hereafter, EVC 8005).

17. Brown, "Profile of Victor McKusick," 35.

18. In the summer of 1964, the Hopkins group also approached the problem of genetic disease from a different direction, focusing not on the stigmata of a known genetic disease but on an indistinct property of the body—mental retardation—that was often present in genetic disease but could have other causes. Their study of mental retardation in the Amish in Lancaster County, Pennsylvania, and Holmes County, Ohio, focused on two groups of approximately equal size but each constituting a closed reproductive pool. They found PKU and a form of mental retardation that seemed to be new, tentatively called "pseudo-mongolism." "While the study of mental retardation in the Lancaster Co. Amish was underway, cases of microcephaly in a parallel 'plain' group in Lancaster Co., the horse-and-buggy Mennonites, came to attention." McKusick quickly found that going out into the field could yield unexpected data about disease and populations. The early work on mental retardation in the Amish is chronicled in McKusick's report to the Kennedy Foundation of 15 November 1964, "Report for Kennedy Foundation on Mental Retardation," box R110D8, VAM. See also "Genetics and the Amish" (editorial), *Journal of the American Medical Association* 189 (1964): 850–51.

19. Later, McKusick kept and bred a dwarf miniature poodle, Vanilla, mating her in a backcross with her sire, apparently with the expectation of producing more dwarf poodles. See letters, 1973, folder M, box R109I9, VAM. Also see my later discussion in the text.

20. McKusick, "Summary of Related Work Already Done," in "Medical Genetic Studies of the Amish," grant application to the National Foundation–March of Dimes for support 1978–80, "National Foundation 1979–80," box R109C16, VAM.

21. For articles reprinted in McKusick's *Medical Genetic Studies of the Amish: Selected Papers Assembled with Commentary* (1978), all page numbers in the text refer to the reprint.

22. Postcard to Victor McKusick, January 1964, EVC 8005.

23. Letter, 13 September 1966, EVC 8005.

24. Letter, September 1969, EVC 8005.

25. Twenty-one pages of hand-written notes describing these sibships and this informant network, probably written in July or August 1963, are in EVC 8005.

26. Other deaths could delay the field research: for example, for sibship 47, "son's child died three days ago. Considered inopportune time to visit" (ibid.).

27. Letter in Amish, box R110D8, VAM.

28. McKusick to Robert Baur, 26 November 1963, with attached hand-written notes in response, EVC 8005. Other correspondence with physicians who treated the Amish is also in this file.

29. McKusick to Clinton Lawrence, 12 September 1963, EVC 8005.

30. Andrew Gale, E. A. Murphy, V. W. McKusick, and V. A. McKusick, "Trip to Lancaster County—17 February 1975," in "Lancaster File," box R110D8, VAM.

31. "4/6/82 Notes on a Trip to Lancaster with Clair Francomano and Guadalupe Gonzales-Rivera (Known as Lupita)," red binder, box R109I18, VAM.

32. McKusick's report to the Kennedy Foundation, 15 November 1964, "Report for Kennedy Foundation on Mental Retardation," box R110D8, VAM.

33. McKusick to Henry S. Wentz, 10 March 1965, EVC 8005.

34. This illustration appeared originally in Theodor Kerckring, *Spicilegium anatomicum*, Amsterdam, 1670. It was reprinted in McKusick 1978.

35. Hereafter, references in the text to McKusick's *Mendelian Inheritance in Man*, in its various editions, are designated MIM.

36. "The Only Way to Cure a Genetic Illness Would Be to Repair the Gene" (interview), *Baltimore Evening Sun*, 3 March 1966, copy in "Biographical Information," VAM; Brown, "Profile of Victor McKusick," 3.

37. McKusick used his work on pedigrees in a 1957 grant application to the NIH for support for a study that could "define, in as much detail as possible, the mode of inheritance, mechanisms of clinical manifestations . . . [and] factors affecting penetrance and expressivity" in cardiovascular disease. The pedigree method, he said, would be "the cornerstone of the program," and the "machinery" for ascertaining

relevant pedigrees was already in place in Baltimore. He would need a nurse for "pedigree tracing, checking hospital records, assistance in clinical testing," part-time secretarial help, and a technician for chemical tests. He also asked for $150 for drawing pedigree charts. "A Study of the Genetic Factor in Cardiovascular Diseases," application for research grant, 20 February 1957 (not in folder), box R109C8, VAM.

38. McKusick (his handwriting), "Pedigree Methods," in "The Pedigree Method in Medical Genetics," box R109C8, VAM.

39. "Amish Grant," box R110D8, VAM.

40. The term *medical tourist* was used by a physician currently treating the Amish for genetic diseases to refer to medical researchers of an earlier era who were "less interested in the health-care needs of the community than they were in the diseases themselves." This is presumably a reference to McKusick and his field workers. D. Holmes Morton, quoted in Susan Q. Stranahan, "Clinic a Lifeline to Children: A Doctor Finds His Calling among the Amish and Mennonites," *Philadelphia Inquirer,* 15 September 1997.

41. The summary referee report is attached to a letter, Paul A. Deming of the National Institute of General Medical Sciences to McKusick, 28 December 1977, for application 1R01 GM 24757-01, filed with grant proposal, "Amish Grant," box R110D8, VAM.

42. McKusick was elected to the NAS in 1973. See large file on his nomination and admission, "NAS," box R109E4, VAM.

43. McKusick to Edward C. Melby, 6 April 1973, and Melby to McKusick, 7 March 1973, correspondence and related materials in folder M, box R109I9, VAM.

44. The hypothesis about sickle cell trait and malaria was explored in field studies in Africa by the Oxford University medical biologist Anthony C. Allison (1954). For a discussion of Allison's field methods, which at one point included infecting some African subjects with malaria, see Kevles 1985, 228–30.

45. McKusick has attributed the growth of human genetics as a clinical discipline to technological change. In a 1980 lecture, for example, he said that "methodologic advances, as always, have made this scientific progress possible. These technical advances have included chromosome banding and somatic cell hybridization and other methods which provided surrogate approaches to the genetics of man." He thus proposed that human genetics moved away from the whole organism (even as he was continuing to work with whole, functioning human beings in his field studies) and into cultured cells in the laboratory. The field had also lost its "exclusive reliance on statistical methods" and become "highly medicalized." McKusick, "Twenty Years in the Study of Genetic Diseases of Man," draft of lecture, 11 January 1980, box 109E3, VAM.

CHAPTER 4. SQUASHED SPIDERS

1. For the standard description of this method of chromosome preparation, the method widely used in this period, see Moorhead et al. 1960. The research leading

to this method was supported by the National Cancer Institute and the work was done in connection with cancer research under way in Philadelphia, at the Wistar Institute of Anatomy and Biology, the departments of pathology and pediatrics at the School of Medicine of the University of Pennsylvania, and the Institute for Cancer Research of Philadelphia. Hungerford and Nowell were the discoverers (later) of the so-called Philadelphia chromosome, the chromosome associated with leukemic cells. On the use of the red bean extract, see J. G. Li and Osgood 1949.

2. Only when "human cytogenetics began to settle down" (apparently, in Patau's construction, in 1960) did it become possible to produce a review volume of lasting value. And so Patau considered John Hamerton's book, which he was reviewing in this complaining mode. "In a field in which many observations are necessarily ambiguous, there remains naturally occasion for disagreement, but one is grateful to be spared those cytogenetic naiveties that mar so many otherwise competent publications" (Patau 1963).

3. "I would suggest that you use the chromosome culture kits which are available from DIFCO . . . the kits provide all that is necessary to obtain chromosome preparations except a blood sample . . . There might be some people at the University of Malaysia at Kuala Lumpur who might be able to help you as regards microphotographic needs." Digamber Borgaonkar in Victor McKusick's lab to Diane Breyers in Negri Sembilan, Malaysia, 1 February 1972, folder B-1970, box R110B13–14, VAM.

4. The first designation refers to a balanced reciprocal translocation between the short arm of a B and the long arm of a D group chromosome ("B group" and "D group" being one of the early classification schemes, now superceded). This particular formula is given as an example in the report of the Chicago Conference, "Standardization in Human Cytogenetics," at the University of Chicago Center for Continuing Education, 3, 4, and 10 September 1966 (Chicago Conference 1966, 4). The second designation is from "A Reciprocal Translocation between a Short Arm of Chromosome No. 2 and the Long Arm of Chromosome 18," in "Chicago Cytogenetics Conference," 11, papers of Curt Stern, American Philosophical Society, Philadelphia (hereafter, APS).

5. The face is apparently under tight genetic control, and many complex syndromes do produce subtle facial signs that can be read, so that the face reflects the chromosomal realignment or the disease gene. In genetic disease groups, those sharing the disease can sometimes resemble each other more than they do their own parents or siblings. Yet the facial differences are often subtle. I have examined photographic guides to these syndromes in various reference works used in medical genetics, and although some affected persons do look very different and obviously malformed in some way, many of the syndromes would not be noticeable to the (my) untrained eye. Seeing genetic disease in the face is familiar knowledge, or craft knowledge, like the knowledge of the natural historian who recognizes subtle dif-

ferences in different plants and animals. It involves some combination of skill at pattern recognition, formal training, and experience with patients.

6. Clarke Fraser, personal communication, 2001.

7. An idiogram is an image of human chromosomes that is a composite from many different cells. A karyotype is from a single cell (see n. 9 below).

8. Chu was at the Biology Division at Oak Ridge National Laboratory and had delivered this paper as part of a symposium in September 1959 at Pennsylvania State University at a meeting of the American Institute of Biological Sciences.

9. In 1960, karyotypes were also known as idiograms, but the conference organizers distinguished between the two: the karyotype is a depiction, either photographic or drawn, of the chromosomes of a *single* cell; the idiogram is a "diagrammatic representation of a karyotype which may be based on measurements of the chromosomes in several or many cells" (Denver Study Group 1960, 1063, lengthy footnote).

10. Hungerford was codiscoverer of the Philadelphia chromosome (chromosome 22), a chromosomal nondisjunction explicitly linked to leukemia (Hungerford 1961).

11. The strong representation of those interested in radiation and cancer at this early meeting reflected the tight links between radiation biology and genetics after 1945.On this general question, see my work on the Atomic Bomb Casualty Commission (Lindee 1994).

12. "Chicago Cytogenetics Conference," typescript, report with supporting documents, in the papers of Curt Stern, MS Coll. 5, APS. This is the source of all the quotations in the following text discussion of the Chicago conference (page numbers as in original report).

13. Minutes, "Chicago Cytogenetics Conference," 3 September 1966, conference papers of Curt Stern, APS.

14. For a review of the Down-Klinefelter relationship slightly later, with a useful bibliography, see Benirschke et al. 1962. The original paper out of Penrose's lab is by Ford et al. (1959).

15. See, for example, Cohen, Shaw, and MacCluer 1966. The bibliography in their paper contains references to other papers on this topic.

16. Quotation from Hungerford et al. 1969, 78. See also estimates of the frequency of this phenomenon in the general population by Court Brown et al. (1966).

17. "Chicago Cytogenetics Conference," 6, APS.

18. Ibid., 11. "The whole subject has now become so vast and complex that research workers, especially those in isolated laboratories, cannot easily keep themselves fully informed about what others are doing. Consequently some mechanism of collection and dissemination of knowledge outside the ordinary channels of scientific publication is probably desirable. There are many difficulties in the establishment of a suitable organization of this kind. A substitute for the Human chromosome Newsletter would require very careful planning."

19. Penrose, 3 September 1966, "Introductory Address," in "Chicago Cytogenetics Conference," APS.

20. Sumner argued in 1982 that it was not enough to merely take advantage of these banding technologies as technological tricks. Their chemical workings needed to be understood, because an understanding of why they produced the effects they did could make chromosome banding more than "a practical tool." It could "provide important insights into a particular level of chromosome organization" (80).

CHAPTER 5. TWO PEAS IN A POD

1. Not all tests of this hypothesis that questionnaires are sufficient to determine zygosity have validated it. In 1977 Louise Carter-Saltzman and Sandra Scarr tested a group of four hundred same-sex adolescent twins ages ten to sixteen for correlation between blood group results and self-reported zygosity. The twins were interviewed rather than sent a questionnaire. Carter-Saltzman and Scarr found that 89 percent of the monozygotic twins were correct in their diagnosis, but only 59 percent of the dizygotic twins were correct.

2. I am hedging a bit here on both the date and the founding of the Behavior Genetics Association. Although the first president of the society was appointed in 1972, comments in several published sources suggest that it "existed" prior to that date, I assume in a more informal mode.

3. Rende, Plomin, and Vandenberg proposed in 1990 that the American professor of education Curtis Merriman also proposed the classic twin method in a 1924 paper drawn from his dissertation. But Merriman did not compare concordance, and other twin researchers generally credited Merriman only with establishing that there were two kinds of twins (identical and fraternal).

4. No twin study ever found 100 percent concordance for schizophrenia in monozygotic twins, so there cannot be a single, determining gene for the disorder. Yet most have found higher concordance in monozygotic than in dizygotic twins, suggesting some hereditary influence.

5. Kallmann (1897–1965) was a student of Ernst Rudin and a fierce hereditarian (Muller-Hill 1998, 28–29). Kallmann was one of the most prominent and earliest promoters of psychiatric genetics. There is a large historical literature on schizophrenia, but I was unable to find a scholarly historical study devoted solely to Kallmann. He had some ties to the Rockefeller Foundation, and the Rockefeller Archives holds some materials relating to his genetics work. For a good sense of his work, see Kallmann 1938, 1953.

6. All page numbers in the text refer to Newman, Freeman, and Holzinger's 1937 book. As Arthur Jensen pointed out in 1974, Newman, Freeman, and Holzinger happened to be at the University of Chicago at the time that Sewall Wright was there. "It was Holzinger who was responsible for all the quantitative treatment of the data. He went about this like a good statistician, but without showing the slightest evidence

that he was aware of anything in the field of quantitative genetics, which at that time was already quite advanced and would have been highly applicable to the excellent data gathered by Newman et al. It should be noted, too, that this study was conducted at the University of Chicago at the time when Sewall Wright, a leader in quantitative genetics, was professor of genetics there; but apparently Wright was not consulted by Newman et al. This is one of those interesting curiousities in the history of science" (Jensen 1974, 25, footnote). This article, by the way, was the one in which Jensen demonstrated some problems with Sir Cyril Burt's twin data.

7. I use the figure two hundred, but in fact there are conflicting estimates of how many such pairs have been found and studied. This is because different authors use different definitions of "at birth" and "separated." Some recognize twins separated as late as age seven or eight as "separated at birth"; others include only those separated within about the first year of life. Some consider twins reared in the same neighborhood by relatives (an aunt and a cousin, for example) to be "separated," and others count as separated only twins who were unaware of each other's existence and/or were separated by great distances and never in social contact during childhood. The lives of twins are as variable as human life stories in general, and twins are not standardized organisms. Deciding what counts as being "separated at birth" is both a nontrivial problem and one resolvable only by consensus, because there is no absolute gold standard for what counts as a shared environment.

8. The Newman, Freeman, and Holzinger study (1937) is an important classic in twin research and is in itself worthy of further study. It was the first major American project focused on twins and was a truly exhaustive review of the literature at the time. The research—with fifty pairs of identical twins, fifty pairs of fraternal, and nineteen separated at birth—coincided with the Chicago World's Fair in 1933, and the researchers, who lived in Chicago, used the lure of free trips to the fair to bring in the twins for examinations.

9. Eliot Slater's 1953 summary of the state of twins research is the primary source for this discussion of early registries. Slater saw registries as a critical part of the twin method and traced their development (7–22).

10. Thomas H. Hunter, at the February 1957 meeting of the Ad Hoc Committee on Studies of Veteran Twins, minutes, in papers of James V. Neel, American Philosophical Society, Philadelphia (hereafter, JVN).

11. The registry was both entirely male and entirely white, and while the maleness of the registry was explicitly discussed as a limitation, the whiteness of the registry was not. African American veterans, when they happened to be twins, were excluded from the study and the registry, this exclusion being stated several times in the planning documents and even in the published literature and never otherwise explained. In a pilot study that examined all twins, one of the identified twin pairs was found to be "Negroid and was excluded from the study on this basis" (Jablon et al. 1967, 135). It must have been the obvious, taken-for-granted sense of this com-

munity that African American twins were somehow different or somehow not relevant to the twin registry, but I have not been able to find this taken-for-granted sense further elaborated.

12. Gilbert Beebe, Bernard M. Cohen, and Seymour Jablon, "Some Observations on Follow-Up Studies of the Veteran Population," at the VA Medical Research Conference, Memphis, Tennessee, 2 December 1953, typescript, JVN.

13. Berkowitz and Santangelo (1999) give a finely detailed historical account of the Medical Follow-Up Agency and its work. This is a contract history, very well done, but the agency deserves the attention of a more distanced scholarly observer.

14. "Report of Progress Under Contract PH43-64-44, Task Order No. 11, Medical Follow-Up and Epidemiological Studies on Veterans," 26 February 1973, in MED: Com. on Epidemiology and Veterans Follow-Up Studies, Mtgs: Agenda 1973 June, National Academy of Sciences (NAS).

15. Beebe, Cohen, and Jablon, "Some Observations," 6, JVN.

16. A fairly elaborate description of the early reasoning is Gilbert Beebe to Philip Owen, 26 June 1946, "Value of Army and Navy Personnel Records in Creating Rosters of Twins," in MED: Com. on Epidemiology and Veterans 1963–68, Follow-Up Studies (MFUA): History, NAS. See also bulletin, "Veteran Twins," 21 February 1957, "Minutes of the First Meeting of the Ad Hoc Committee on Veteran Twins," JVN.

17. "Veteran Twins," JVN.

18. "Medical Follow Up Based on Military Experience," 7, JVN. I think Gilbert Beebe wrote this document, and I think the date is 1957, but neither an author nor a date is provided so I am inferring both from the content and the context.

19. Ibid., 8.

20. "The Role of Heredity in Disease as Determined by Studies on Veteran Twins," draft, 29 June 1956, presumably written by Neel, JVN.

21. Ibid.

22. "Veteran Twins," JVN.

23. National Research Council, Division of Medical Sciences, Follow-Up Agency, "The Role of Heredity in Disease as Determined by Studies on Veteran Twins," draft protocol, 14 December 1956, JVN.

24. Bernard M. Cohen, draft letter to State Registrars, 25 June 1956, JVN.

25. "When dermatoglyphic methods were first introduced for the diagnosis of twin zygosity around 1930 they filled a great need," Gordon Allen of the National Institute of Mental Health proposed, but blood typing made them appear less valuable. "Perhaps as a reaction against the difficulties of serological twin diagnosis in large series of twins, there is now some return to subjective methods. Even mailed questionnaires are thought to be sufficiently reliable for some purposes" (G. Allen 1968, 360).

26. The utility of eye and hair color was also culturally specific. As a group of researchers on Taiwanese twins reported in 1999, "It is not surprising that items use-

ful in Western countries such as eye color and hair color in fact were useless in this study" (Chen et al. 1999). See also "The Role of Heredity in Disease," JVN.

27. The students were R. D. Budson, J. D. Burke, T. W. Chaffee, H. R. Netzer, R. G. Wonacott, P. A. Yalowitz, and Jean Hurchalla. They delivered, Neel said, "a conscientious performance that brooked no obstacles." Typescript for 1967 paper, 5a, JVN.

28. Parisi and Di Bacco (1968) found that fingerprints provided a fairly high general probability (.86) of correct diagnosis, though they also described many of the practical difficulties of using fingerprints. All use of fingerprints requires scoring for similarity—"the probability of monozygosity is indirectly proportional to the difference in the co-twins ridge counts."

29. Jablon, in minutes, MED: Com. on Epidemiology and Veterans Follow-Up Studies, no. 2, 28 March 1966, 7, NAS.

30. Dunn, minutes, Ad Hoc Committee on Studies of Veteran Twins, Bulletin, Veteran Twins, 21 February 1957, 1–8, quotation from p. 6, NAS.

31. Proposals in which the twin status of the veterans was almost irrelevant were rather common in the early years, possibly because it was not yet clear to the biomedical community how the twins should be used. Minutes, Com. on Epidemiology and Veterans Follow-Up Studies, no. 9–11, April 1972, 6; referring to a study by Ray Rosenman of the Harold Brunn Institute for Cardiovascular Research, in MED: Com. on Epidemiology and Veterans Follow-Up Studies, Mtgs: Minutes, 1972 April, NAS.

32. Minutes, in MED: Com. on Epidemiology and Veterans Follow-Up Studies, no. 3, 27 April 1967, 9–10, NAS.

33. Ibid., 11.

34. Ibid., 12.

35. "Guidelines for Access to the Twin Registry, draft, 1 May 1968, Tab. III, in MED: Com. on Epidemiology and Veterans Follow-Up Studies, Mtgs: Agenda, 1968 May, NAS.

36. "Status of Continuing Projects," 20 December 1972, in MED: Com. on Epidemiology and Veterans Follow-Up Studies, Mtgs: Agenda attachments IV–VIII, 1973 January, NAS.

37. As stated in letter of 21 September 1971 to twins, signed by Charles L. Dunham, in MED: Com. on Epidemiology and Veterans Follow-Up Studies, NAS.

38. Zdenek Hrubec to Subcommittee on Twins, 21 September 1971, draft of letter to members of the NRC Twin Registry, in MED: Com. on Epidemiology and Veterans Follow-Up Studies, NAS.

39. The role of patient support groups in contemporary genetics research is the subject of recent work by Taussig (2005), Taussig, Rapp, and Heath (2002), and Rapp (2003), who show that families with genetic disease are in some cases equivalent to full participants in the research.

40. Zdenek Hrubec to Paul Taubman, 20 October 1972, in MED: Com. on Epi-

demiology and Veterans Follow-Up Studies, Mtgs: Agenda and attachments I and II, 1973 January, NAS.

41. All page numbers in the text are from Paul Taubman, "A Proposal to Study the Importance of Nature and Nurture on Earnings, Occupational Mobility, Healthiness, and Family Size Using the NAS-NRC Twin Sample," 1972, in MED: Com. on Epidemiology and Veteran's Follow-Up Studies, Mtgs: 1973 January, NAS.

42. See specifically Robert W. Morse, chair of the NRC's Board on Human Resources, to Zdenek Hrubec, 24 August 1972, in MED: Com. on Epidemiology and Veterans Follow-Up Studies, Mtgs: Agenda and attachments I and III, 1973, NAS. In this letter Morse explicitly urges Hrubec to take Taubman's proposal seriously, partly because it involved a social scientist exploiting a medical resource. It also involved a person who was the chair of Morse's board's Panel on the Benefits of Higher Education.

43. This statistic is less puzzling if one explains alcoholism as a product of social expectations and practices. Alcohol consumption is more acceptable for young men, who are the group at highest risk of alcoholism, and women may be better able to conceal their alcohol abuse.

44. See Peele's website, www.peele.net (accessed 4 January 2004); Peele 1986.

45. In this same address, Scarr described being physically threatened, once with murder, by students enraged at her studies of black children adopted by white families. She reported an incident when students stormed the stage after a talk at the University of Minnesota by Arthur Jensen (Scarr 1987, 224–25).

46. These adjectives are from the Harrison G. Gough Adjective Check List (1952), one of the tests given to veteran twins as part of a study by Paul Rosenman on "genetic aspects of type-A behavior." Subjects were told to "read quickly and blacken in the circle beside each one you would consider to be self-descriptive. Do not worry about duplications, contradictions and so forth. Work quickly and do not spend too much time on any one adjective. Try to be frank and fill in the circles for the adjectives which describe you as you really are, not as you would like to be." A copy of this test is filed in MED: Com. on Epidemiology and Veterans Follow-Up Studies, Subcom. on Twins Studies, Mtgs (teleconference), 1972 December, NAS.

CHAPTER 6. JEWISH GENES

1. Penicillin became available in American pharmacies in April 1945.

2. In the popular press, see Jon Katz, "My Other Children," *Glamour*, March 1994; "Healthy Baby Is Born after Test for Deadly Gene," *New York Times*, 17 January 1994; Gina Kolata, "Nightmare or the Dream of a New Era in Genetics?" *New York Times*, 7 December 1993; Jared Diamond, "Curse and Blessing of the Ghetto," *Discover*, March 1991; and many more.

3. Riley, "A Short History of Familial Dysautonomia" (banquet, Dysautonomia Foundation, 19 November 1970).

4. Interview with Lenore Roseman, New York, 28 October 1997.

5. A twelve-year-old boy whose histamine response flared weakly—registering as neither a normal response nor an FD response—they identified as suffering from a "dysautonomic defect of very mild nature." Another boy did not have any dysautonomic defect, they concluded. "His response to histamine is completely normal. Because of this, the patient's record was reviewed critically and the patient examined by us. He was found to have congenital heart disease" rather than dysautonomia (Smith and Dancis 1962).

6. They showed that the subnormal response to intradermal histamine could be present also in children with a different disease, pheochromocytoma (Smith and Dancis 1964b).

7. This list is from Axelrod's *Manual of Comprehensive Care* (1997a).

8. Interview with Axelrod, New York University Medical Center, 27 October 1997.

9. Quoted in Stefanie Dell'Aringa, "Dr. Axelrod Brings Hope to Patients with Rare Disease," *American Academy of Pediatrics News*, May 1997.

10. "Dysautonomia Day at New York University Medical Center," *Genesis* 5, no. 3 (1993). *Genesis* is the newsletter of the Genetics Network of the Empire State, New York State Department of Health. See also Felicia B. Axelrod, "Report, Dysautonomia Day" (typescript, June 1994), obtained from Axelrod, October 1998.

11. "The Dysautonomia Treatment and Evaluation Center at NYU Medical Center," *Genesis* 2, no. 2 (1989).

12. Axelrod, "Report, Dysautonomia Day," 1994.

13. Felicia Axelrod, "Report, Dysautonomia Day, June 1997" (typescript), from Axelrod, October 1998.

14. Susan A. Slaugenhaupt and James F. Gusella, "The Search for DYS; The Search for George: A brief description of the molecular genetic methods we are using to identify the familial dysautonomia gene, DYS, explained with the help of George, a mysterious person with six toes on his left foot" (Dysautonomia Day Report, 1997, typescript, Dysautonomia Foundation), from Felicia Axelrod, October 1998. Dysautonomia Day is a meeting of patients, their families, and the scientists and physicians who work on FD. The program includes scientific reports on the status of the search for the gene and practical advice on managing problems.

15. Ibid.

16. Susan Slaugenhaupt, "Dysautonomia Day 1999" (typescript), from Axelrod.

17. Lynn D. Frohnmayer to Joan O. Weiss, 29 September 1994, letter describing the development of the Fanconi Anemia Research Foundation (reprinted as appendix C in J. O. Weiss and Mackta 1996, 133–36).

18. This account of the gene hunt is from a profile of the Dysautonomia Foundation by Gina Kolata (1996). The 1997 report explaining the gene hunt to families affected by FD listed three candidates for the dysautonomia gene. One was called

CG-1 ("CG stands for Cool Gene"), which seemed to be associated with a protein in the cytoskeleton of the cell: "FD? Disruption of the cytoskeleton can cause cell death during development." CG-2 and CG-6 were also candidates, though of unknown function. CG-6 "does not look like anything in the database," and CG-2 "looks like a worm gene." Slaugenhaupt and Gusella, "Search for DYS," Dysautonomia Foundation.

19. Dysautonomia Foundation, "Dysautonomia Foundation Inc.—History, Programs, Accomplishments" (press release, New York, 1997).

20. Slaugenhaupt and Gusella, "Search for DYS," Dysautonomia Foundation.

21. Axelrod, "Report, Dysautonomia Day," June 1993.

22. Interview with Robert Dancis, NYU Medical Center, 28 October 1997.

23. Interview with Lenore Roseman and Maryon Weill, Dysautonomia Foundation Offices, New York, 28 October 1997. Roseman and Weill do a great deal of emotional work relating to FD. Weill "cries every time we get a new patient." Both women report that "parents turn to us when they get angry or upset." They also organize each year's Dysautonomia Day, and they try to raise money to fund research and help for Axelrod at the center. They have become part of the social and emotional network built around FD. Weill has a close family member, a nephew, who has FD, but Roseman has no familial connection to the disease.

24. Amy Rutkin, quoted in *Alliance Alert* (a publication of the Alliance of Genetic Support Groups), November 1997.

CHAPTER 7. CONCLUSIONS

1. "Gene Therapy Patents," *Cancer Gene Therapy* 1, no. 3 (1994): 227–34.

2. Broder and Anderson's comments in transcript of proceedings (President's Cancer Panel 1990, 6, 44).

3. Press release, University of Pennsylvania Medical Center, 7 December 1992.

4. Biographical profile, James M. Wilson, released December 1992 by the University of Pennsylvania Department of Public Affairs (Carey 1994, 84).

5. See comments by Paul Gelsinger, Jesse's father, at www.geneletter.org/05–01–00/features/gelsinger1.html.

6. Engineering the Human Germline Symposium, summary report, June 1998, www.ess.ucla.edu:80/hige/report.html.

7. "Trials and Tribulations" (editorial), *Nature Genetics* 24, no. 3 (2000).

8. This presentation is in the biographical sketch provided to journalists when Wilson was appointed head of the new Institute for Human Gene Therapy at Penn in 1992. I requested data about the creation of the institute from the press office there, long before the death of the teen research subject Jesse Gelsinger in September 1999.

9. Engineering the Human Germline Symposium, summary report.

10. Victoria Odesina, People's Genome Celebration, Washington, D.C., 9 June 2001.

11. Paul Jacobs, "A Father's Mission," *San Jose Mercury News*, 31 December 2000.

12. Francis Collins, People's Genome Celebration, Washington, D.C., 9 June 2001.

13. "The People's Science," *Wellcome News*, no. 26, Q1 (2001): 6–7.

ESSAY ON SOURCES

The published literature on genetic disease is diverse, but serious historical attention is rare and none of what I have read takes exactly the approach I have chosen. I am indebted to many participant historians who have analyzed their fields, thought about historical change, and identified key questions and problems. Barton Childs's sustained and deep attention to genetic medicine constitutes the model example of the genre (Childs 1999). Victor McKusick wrote histories of genetic disease and genetic medicine at several stages in his long career, and his work provides insight into his own changing views of the important issues and problems and into the general development of the field (McKusick 1969, 1981, 1992, 1996). Essays, lectures, and historical papers by James V. Neel, Irving Gottesman, Felicia Axelrod, Lionel Penrose, Alfred Knudson, Joseph Fraumeni, T. C. Hsu, Ken Weiss, and many other technical experts were crucial resources. These accounts rarely focused on the questions I was tracking—most historians would think of them as primary documents. But for historians working on the period after 1945, such accounts are often best seen as a mixture of primary and secondary sources. They provide insight into a community's sense of its own history, and in that sense they are primary sources, a focus of explicit study. But they also provide rough empirical maps of important people and events, of dates, technologies, institutions, and theoretical problems. They are the first draft, the starting point, and without them a study of this scope on such recent materials would be impossible.

Persons whose families are affected have also produced studies of genetic disease, as have physicians who have worked with genetic disease patients. These first-person accounts generally emphasize the promise of medical intervention, but they can also provide insights into the role and place of folk and personal knowledge in the systems built around the interpretation of genetic disease. Daniel Pollen's 1993 study of the genetics of Alzheimer disease, *Hannah's Heirs*, for example, documents the efforts of a single large family to create an empirical trail that could help family members assess their own risks. The family members were descendants of a Ukrainian woman who was born in 1844 and died in the late 1890s. Her death may have resulted from an asthma attack, but she also suffered from dementia beginning in middle age. Eight of her nine children emigrated to the United States, and many of them also began to lose short-term memory in their forties or early fifties. In the 1950s, several members of this family, including a physician who was a grandson of Hannah, began collecting data on Hannah's disease and its appearance in the family tree. They created a textual record in the hope that they could learn more about their own risks, and their work, which continued for decades, was informed by a fairly sophisticated folk genetics. Family members came to believe that the disease was becoming less virulent in later generations. "They assumed that whatever factor caused the disease would blend with new genes coming into the family [and] intuitively made the same assumptions regarding a blending of 'factors' that pre-Mendelian breeders had made earlier," Pollen notes (36). Despite his clear status as a promotional participant-historian, Pollen is remarkably attuned to the complex networks around Alzheimer's as experienced disease and as a focus of scientific research. He notices folk and emotional knowledge, the politics of body tissue distribution, and the historical contextuality of technical knowledge.

Historian Alice Wexler's 1995 memoir of her own family's struggle with Huntington disease (HD), *Mapping Fate*, is similarly revealing. Wexler combines an account of personal experience with a perceptive exploration of contemporary genomics and its impact on families. She writes as a person at risk for a devastating genetic disease and as a family member who has experienced its effects. Her mother died of HD, and the Wexlers—Alice, and her sister, Nancy, and their father, Milton—played a critical role in the research that found the gene. Alice Wexler's account provides insight into the motivations and complexities of this personal and scientific engagement. A marker for the Huntington gene was identified in 1983, and the gene itself was finally found ten years later. Because the disease appears only in middle age, HD testing has been an important focus for debates about presymptomatic testing and testing for future disease risk. The disease appears in middle age but can be detected in a fetus: to possess this gene is to possess a disease decades before it can be expected to have any effects.

The search for the HD gene engaged many nonscientists. Alice Wexler has a Ph.D. in history, and in her study she notices the nonscientists in her family and in her social network who collaborated with the geneticists to organize research efforts. But collaborators also included the people living around Lake Maracaibo in Venezuela, families in which the HD gene is so common that some children are born with two copies of this dominant, fatal gene. These Venezuelan families, as Wexler notes, knew a great deal about the disease. They could gauge its progress, diagnose themselves, and assess their own stages of debility. They could see the earliest signs of disease in another family member. They had seen enough bodies affected by HD to categorize its progress with exquisite specificity (Wexler 1995).

Historians' treatments of genetic disease are relatively limited. There is a large and growing literature on molecular biology, bioinformatics, molecular genetics, biotechnology, cancer genetics, and other high-tech or high-theory endeavors (de Chadarevian 2002; de Chadarevian and Kamminga 1998; Fujimura 1996; Gottweis 1998; Hilgartner 1998; Kay 1993, 2000; Keller 2000; Morange 1998; Rabinow 1996, 1999). A picture of the development of the molecular sciences is gradually coming into focus in these diverse works. It suggests that the forces shaping human genetics in the 1960s, including corporate interest and the rise of computer sciences, were also implicated in genetics more broadly.

There is also a large and increasingly sophisticated literature on eugenics both before and after 1945, in many different national contexts (Adams 1989a, 1989b; S. F. Weiss 1987; Proctor 1995; Kevles 1985; Paul 1995, 1998, 1999a, 1999b). Such studies explore the variation in eugenic policies, practices, and beliefs in different settings and track the relationships between scientific and social values. Of particular interest is the work of Diane Paul, who is both a political scientist and historian and has explored the rise of human genetics, the relationships between human genetics and eugenics, and the shifting effects of new technologies and theories. With a consistent interest in the ironies of technosocial change, Paul has interrogated critically some of the most treasured narratives of postwar human genetics. For example, in a series of papers and reports, sometimes working with Paul Edelson, she has explored one of the great success stories of public health genetics, the low-phenylalanine diet for the control of phenylketonuria (PKU). As Paul shows, PKU has been a favorite disease for critics of genetic determinism, because diet can prevent mental retardation. The disease provided evidence in the nature-nurture debate, weighing in resolutely on the side of nurture. At the same time, PKU has been a favorite disease of promoters of genetic testing and screening, because it can demonstrate the benevolent potential of the acquisition of genetic information. "The PKU story is infinitely plastic, employed by both celebrants and skeptics of genetic medicine," she notes. Neither side in this debate, however, has been attentive to the real difficulties of dietary management. The low-phenylalanine diet creates profound hardship in affected families, and while most consequences of the disease are ameliorated, they are not entirely prevented. As Paul suggests, the uncertainties and ambiguities of PKU testing provide insight into the problems posed by genetic testing more generally (Paul 1995, 1998, 1999a, 1999b).

From a related perspective, Edward Yoxen's unpretentious 1982 paper, still very much worth reading, poses the critical question of "why we isolate or delineate certain phenomena for analysis, why we say that they constitute diseases and why we seek to explain their nature and cause in genetical terms" (144). He suggests that all scientific and medical knowledge has a sociology and that this does not require the dismissal of medical knowledge as a phantom. Genetic models of disease can be grounded reliably in "material reality" while also involving choices about research programs, technologies, and evidence. Genetic research depends on "certain judgments of resemblance and dissimilarity," and these judgments are culturally negotiated. Yoxen's paper sketches some critical questions that have informed my work.

Professional historians of medicine have also looked at genetic disease. Particularly relevant is Keith Wailoo's analysis of the historical development of hematology and the clinical management of sickle cell anemia. In *Drawing Blood* (1997), his survey of the development of American hematology since the late nineteenth century, he interrogates the relationships of patients, doctors, diseases, and technologies, exploring shifting professional aspirations, corporate concerns, and technical capabilities in the interpretation and management of hematological conditions. His related study of sickle cell disease in Memphis, *Dying in the City of the Blues* (2001), explores the social invisibility of the pain of patients with sickle cell disease. He considers the potency of racial stereotypes in medical diagnosis and the power of social circumstances to dictate treatment. Wailoo's work reflects a strong tradition in the history of medicine that is focused on contested authority and community notions of disease, morality, and illness, a framing pioneered in the 1960s by Rosenberg and subsequently the basis of a large corpus of historical work.

Rosenberg himself, in a 2002 lecture (unpublished) at Harvard Medical School, proposed that the "utopian vision" of knowledge and control that infuses contemporary genomic medicine is expected by its proponents to "dissolve the traditional oppositions whose relationships had defined medicine in previous eras: Art as opposed to science, mind versus body, idiosyncrasy as opposed to general pattern, reductionistic medicine as opposed to holistic inclusiveness." As he noted, this vision is premised on the idea that if there were just enough scientific information, enough technical information, there would no longer be a need for the "art of medicine." The individual specificities of each patient could be "spelled out in molecular terms. Randomness, educated intuition and indeterminacy would be replaced by knowledge, control and prediction." This vision offers economic, administrative, and intellectual rewards, and immanent in this dream is the goal of banishing pain and perhaps even death and "of reducing emotions and behavior to the working out of innate neurochemical and neuroanatomical realities." His account interrogates some of the historical narratives I examine in this study. For example, in the fall of 1952 the German-born psychiatrist Franz J. Kallmann proposed in an after-dinner lecture to the American Society of Human Genetics that the study of heredity could provide an "antidote to the excitement of rebellious anguish, revolutionary destructiveness or supernatural symbolism in frustrated people." Genetic knowledge, he said, might not prevent people from experiencing "excess emotions," but it could provide a constructive guide to "health, sanity and social order" (Kallmann 1952, 244–45). Kallmann's vision of human genetics as a science of rational social order that could overcome the dangerous emotions associated with mysticism or rebellion is precisely the vision Rosenberg's critique explores.

My approach has also been shaped by recent work in the anthropology of science and medicine. Rayna Rapp's ongoing analyses of genomic cultures, family stories, and bodily differences are located at the same point of contact where most of my own analysis lodges. In her ethnography of fetal testing, Rapp explores the views of genetic counselors, prospective parents, and physicians as they negotiate their way through the information provided by amniocentesis. She follows the differences expressed through class and race, looks at the stories told about disability and genetic disease by counselors and others, and looks with particular sensitivity at those who refused the technical imperative and chose either to avoid amniocentesis entirely or to bear a child with a detectable bodily difference (Rapp 2000). Fetal testing is the clinical front line of genetic medicine, the place where the most effective technology for the control of genetic disease is deployed, and the site of contested definitions of family, body, child, and relationship that are evocatively tracked in Rapp's work.

In ways that closely intersect with my interests, Karen-Sue Taussig has grouped the "lay" communities built around genomics into three categories: those who are containers of DNA or who hold knowledge of interest to molecular geneticists (e.g., family medical histories and pedigrees), those who produce genetic knowledge (scientists), and those who consume it. Taussig explores the social processes operating at the intersection of people's bodies, corporate interests, medical activism, and biomedical research agendas. Her work focuses on the contemporary state of genomic medicine, but her concerns are directly relevant to my own (Taussig 2005).

In a related project, Rapp, with Taussig and Deborah Heath, has been exploring the production, circulation, and uses of genetic knowledge among research scientists who discover genes, physicians treating people whose conditions are the result of those genes, and individuals and families in support groups living out the experiences of genetic disorders. They have looked at shifting notions of bodily control among groups such as the Little People of America and at the involvement of parents and family members with technical research, sometimes as teachers of medical students or experts who achieve authorship on scientific papers. The concerns that drive their anthropological work, concerns about the joint production of genomic practice and theory, have been critical to my historical work as well.

As my project focused increasingly on the ways that knowledge moves from the research subject to the scientific text and on other kinds of knowledge that have some relationship to technical knowledge, I found that my interests were also intersecting with literature that has no obvious relevance to genetic disease. My interests became cognate with those of the growing number of scholars interested in experimental organisms. Flies, dogs, mice, sheep, scallops, and other creatures have acquired a high visibility in science studies in the last decade or so. The literature on these organisms that do not speak in human language proved to be a compelling guide, for it was the prevailing premise of passivity, a premise challenged by the animal work, that led me to notice just how active human subjects were in many of the texts I examined.

The issues raised by animal models and technologies often begin with the question of the passive or active agent. Robert Kohler (1994), in his study of T. H. Morgan and the use of the common fruit fly, *Drosophila melanogaster*, as a model for genetic research, chose the simple strategy of paying attention to the fly and its habits. Fruit flies, in his account, more or less made themselves a resource for the scientists. They behaved and reproduced and mutated in ways that facilitated genetic research. At the same time, the members of Morgan's fly group remade themselves as shrewd observers of *Drosophila*. Through the practice of long observation, they acquired the ability to see the flies' most minute variation. The flies and the geneticists, in his story, are working together. Dan Todes has similarly invoked Pavlov's "dog-technologies" as a critical part of the forces and relations of production in the laboratory (Todes 2002). In both cases, these scholars were exploring specialized settings (Morgan's lab, Pavlov's lab) with strong lead investigators and a distinctive culture of production. In such a setting, the biological or biosocial properties of the experimental organism (the "nature" of the living thing) were modified and manipulated to meet the needs of the laboratory. The dogs' anatomies were surgically altered to collect gastric juices. The flies' reproduction was experimentally altered to track genes. At the same time, these accounts suggest, both dogs and flies could and did "resist," producing unexpected mutations, responding differently to protocols, or even, in the case of the dogs, dying in the middle of an important experiment.

One of the delights of scholarship is the opportunity to extrapolate from the ideas of others and apply them to one's own problems. I am indebted to all these scholars for their insights.

BIBLIOGRAPHY

Acosta, Phyllis. 1977. *Diet management of PKU for infants and preschool children.* Washington, DC: U.S. Department of Health, Education and Welfare, no. 77-5209.

Adams, Mark B. 1989a The politics of human heredity in the USSR, 1920–1940. *Genome* 31 (2): 879–884.

———. 1989b. *The wellborn science: Eugenics in Germany, France, Brazil and Russia.* Oxford: Oxford University Press.

Aguayo, A. J., C. P. Nair, and G. M. Bray. 1971. Peripheral nerve abnormalities in the Riley-Day syndrome: Findings in a sural nerve biopsy. *Archives of Neurology* 24:106–116.

Allen, Garland E. 2001. Is a new eugenics afoot? *Science* 294 (5540): 59–61.

Allen, Gordon. 1968. Diagnostic efficiency of fingerprint and blood group differences in a series of twins. *Acta Geneticae Medicae et Gemellologiae* 17 (2): 359–374.

Allen, Richard J. 1960. The detection and diagnosis of phenylketonuria. *American Journal of Public Health* 50 (11): 1662–1666.

Alliance of Genetic Support Groups. 1995. *Directory of national genetic voluntary organizations.* Chevy Chase, MD: The Alliance.

Allison, A. C. 1954. Protection afforded by sickle cell trait against subtertial malarial infection. *British Medical Journal* 1:290–294.

Aly, Gotz, Peter Chroust, and Christian Pross. 1994. *Cleansing the fatherland: Nazi medicine and racial hygiene,* trans. Belinda Cooper; foreword by Michael H. Kater. Baltimore: Johns Hopkins University Press.

American Society of Human Genetics. 1995. Report. *American Journal of Human Genetics* 57:1233–1241.

Anderson, John A., and Kenneth F. Swaiman. 1966. *Phenylketonuria and allied metabolic diseases*. Washington, DC: U.S. Children's Bureau.

Anderson, S. L., R. Coli, I. W. Daly, E. A. Kichula, M. J. Rork, S. A. Volpi, J. Ekstein, and B. Y. Rubin. 2001. Familial dysautonomia is caused by mutations of the IKAP gene. *American Journal of Human Genetics* 68 (3): 753–758.

Andrews, Lori B., Jane E. Fullarton, Neil A. Holtzman, and Arno G. Motulsky (Committee on Assessing Genetic Risks, Institute of Medicine), eds. 1994. *Assessing genetic risks: Implications for health and social policy*. Washington, DC: National Academy Press.

Armstrong, M. D., and F. H. Tyler. 1955. Studies on phenylketonuria: I. Restriction of phenylalanine intake in phenylketonuria. *Journal of Clinical Investigation* 34:565–580.

Arnstein, Shirley Young. 2000. *No tears: Living with familial dysautonomia*. Phoenix, AZ: Via Press.

Aronowitz, Robert. 1998. *Making sense of illness: Science, society and disease*. Cambridge: Cambridge University Press.

Axelrod, Felicia. 1997a. *Familial dysautonomia: Manual of comprehensive care*. New York: Dysautonomia Foundation.

———. 1997b. Familial dysautonomia. In *Clinical autonomic disorders*, ed. P. A. Low, 2nd ed., 525–535. Philadelphia: Lippincott-Raven.

Axelrod, Felicia, Hedi L. Leistner, and Robert Porges. 1974. Breech presentation among infants with familial dysautonomia. *Journal of Pediatrics* 84 (1): 107–109.

Barr, M. L., and E. G. Bertram. 1949. A morphological distinction between neurones of the male and female, and the behavior of the nucleolar satellite during accelerated nucleoprotein synthesis. *Nature* 163:676.

Bender, M. A., and M. A. Kastenbaum. 1969. Statistical analysis of the normal human karyotype. *American Journal of Human Genetics* 21 (4): 322–351.

Benirschke, K., Lydia Brownhill, D. Hoefnagel, and F. H. Allen Jr. 1962. Langdon Down anomaly (mongolism) with 21/21 translocation and Klinefelter's syndrome in the same sibship. *Cytogenetics* 1:75–89.

Bergen, A. A., A. S. Plomp, E. J. Schuurman, S. Terry, M. Breuning, H. Dauwerse, J. Swart, M. Kool, S. van Soest, F. Baas, J. B. ten Brink, and P. T. de Jong. 2000. Mutations in ABCC6 cause pseudoxanthoma elasticum. *Nature Genetics* 25 (2): 228–231.

Berkowitz, Edward D., and Mark J. Santangelo. 1999. *The Medical Follow-Up Agency: The first fifty years, 1946–1996*. Washington, DC: National Academy Press.

Bessman, S. P. 1968. PKU—some skepticism. *New England Journal of Medicine* 278 (21): 1176–1177.

Bessman, Samuel, and Judith Swazey. 1971. Phenylketonuria: A study of biomedical legislation. In *Human aspects of biomedical innovation*, ed. Everett Mendelsohn, Judith Swazey, and Irene Taviss, 49–76. Cambridge: Harvard University Press.

Bevis, D. C. A. 1952. The antenatal prediction of haemolytic disease of the newborn. *Lancet* 1:395–398.

Bickel, H. 1996. The first treatment of phenylketonuria. *European Journal of Pediatrics* 155 (July, suppl. 1): S2–S3.

Bickel, H. Gerrard, and E. M. Hickmans. 1953. Influence of phenylalanine intake on phenylketonuria. *Lancet* 2:812–813.

Blum, K., E. P. Noble, P. J. Sheridan, A. Montgomery, T. Ritchie, P. Jagadeeswaran, H. Nogami, A. H. Briggs, and J. B. Cohn. 1990. Allelic association of human dopamine D receptor gene in alcoholism. *Journal of the American Medical Association* 263 (15): 2055–2060.

Blumenfeld, A., S. A. Slaugenhaupt, F. B. Axelrod, D. E. Lucente, C. Maayan, C. B. Liebert, L. J. Ozelius, J. A. Trofatter, J. L. Haines, X. O. Breakefield, et al. 1993. Localization of the gene for familial dysautonomia on chromosome 9 and definition of DNA markers for genetic diagnosis. *Nature Genetics* 4:160–164.

Breo, Dennis L. 1994. The cancer revolution—from black box to genetic disease. *Journal of the American Medical Association* 217 (18): 1452–1454.

Bridges, T. J., J. L. Pool, and C. M. Riley. 1949. Central autonomic dysfunction with defective lacrimation: II. Preliminary report of effect of neurosurgery in one case. *Pediatrics* 3 (4): 479–481.

Brumberg, Joan Jacobs. 1988. *Fasting girls: The emergence of anorexia nervosa as a modern disease*. Cambridge: Harvard University Press.

Brunt, P. W., and Victor A. McKusick. 1970. Familial dysautonomia: A report of genetic and clinical studies, with a review of the literature. *Medicine* 49 (5): 343–374.

Carey, John. 1994. The next wonder drug may not be a drug. *Business Week*, 9 May.

Carter-Saltzman, Louise, and Sandra Scarr. 1977. MZ or DZ? Only your blood grouping laboratory knows for sure. *Behavior Genetics* 7 (4): 273–280.

Caspersson, T., S. Farber, G. Foley, J. Kudynoski, E. Modest, E. Simonsson, U. Wagh, and L. Zech. 1968. Chemical differentiation along metaphase chromosomes. *Experimental Cell Research* 49 (1): 219–222.

Cattell, Jacques, ed. 1955. *American men of science*, 9th ed. Lancaster, PA: Science Press; New York: R. R. Bowker.

———. 1972. *American men and women of science: Formerly American men of science; a biographical directory founded in 1906*, 12th ed. Lancaster, PA: Science Press; New York: R. R. Bowker.

Cederlof, R., L. Friberg, E. Jonsson, and L. Kaij. 1961. Studies on similarity diagnosis in twins with the aid of mailed questionnaires. *Acta Genetica* 11:338–362.

Centerwall, Willard, and Siegfried Centerwall. 1965. *Phenylketonuria: An inherited metabolic disorder associated with mental retardation.* Washington, DC: Children's Bureau, U.S. Department of Health, Education and Welfare.

Centerwall, Willard R., Robert F. Chinnock, and Albert Pusavat. 1960. Phenylketonuria: Screening programs and testing methods. *American Journal of Public Health* 50 (11): 1667–1677.

Chadwick, B. P., M. Leyne, S. Gill, C. B. Liebert, J. Mull, E. Mezey, C. M. Robbins, H. W. Pinkett, I. Makalowska, C. Maayan, A. Blumenfeld, F. B. Axelrod, M. Brownstein, J. F. Gusella, and S. A. Slaugenhaupt. 2000. Cloning, mapping and expression of a novel brain-specific transcript in the familial dysautonomia candidate region on chromosome 9q31. *Mammalian Genome* 11 (1): 81–83.

Chen, W. J., H. W. Chang, M. Z. Wu, C. C. Lin, C. Chang, Y. N. Chiu, W. T. Soong. 1999. Diagnosis of zygosity by questionnaire and polymarker polymerase chain reaction in young twins. *Behavior Genetics* 29 (2): 115–123.

Chicago Conference: Standardization in Human Cytogenetics. 1966. Report. *Birth Defects Original Articles Series* 2 (2).

Childs, Barton. 1970. Sir Archibald Garrod's conception of chemical individuality: A modern appreciation. *New England Journal of Medicine* 282 (2): 71–77.

——. 1999. *Genetic medicine: A logic of disease.* Baltimore: Johns Hopkins University Press.

Chu, Ernest H. Y. 1960a. The chromosome complements of human somatic cells. In *Symposium on cytology and cell culture genetics of man,* ed. Gordon Allen, 97–103. Bethesda, MD: National Institutes of Health.

——. 1960b. The chromosome complements of the human somatic cell. *American Journal of Human Genetics* 12 (1): 97–103.

Clark, T., and M. Hughes. 1992. *Sickle cell mutual help groups: African Americans supporting one another.* Chapel Hill: Psychological Research Division, University of North Carolina.

Clause, Bonnie Tocher. 1993. The Wistar rat as a right choice: Establishing mammalian standards and the ideal of a standardized mammal. *Journal of the History of Biology* 26:333.

Clayson, David, Wooster Welton, and Felicia Axelrod. 1980. Personality development and familial dysautonomia. *Pediatrics* 65 (2): 269–274.

Clifford, C. A., and J. L. Hopper. 1986. The Australian NHMRC Twin Registry: A source for the Australian scientific community. *Medical Journal of Australia* 145:63–65.

Cockburn, W. C., B. Hobson, J. W. Lightbown, J. Lyng, and D. Magrath. 1992. The international contribution to the standardization of biological substances: III. Biological standardization and the World Health Organization 1947–1990. *Biologicals* 20 (1): 1–10.

Cohen, M. M., Margery Shaw, and Jean W. MacCluer. 1966. Racial difference in the length of the human Y chromosome. *Cytogenetics* 5:34–52.

Collins, Francis. 1998. Foreword. In *Inherited susceptibility to cancer: Clinical, predictive and ethical perspectives,* ed. William D. Foulkes and Shirley V. Hodgson. Cambridge: Cambridge University Press.

Committee for the Study of Inborn Errors of Metabolism. 1975. *Genetic screening: Programs, principles and research.* Washington, DC: National Academy of Sciences.

Comuzzi, A. G., and D. B. Allison. 1998. The search for human obesity genes. *Science* 280:1374–1377.

Cook-Deegan, Robert Mullan. 1995. *Gene wars: Science, politics and the Human Genome Project.* New York: W. W. Norton.

Court Brown, W. M., K. E. Buckton, P. A. Jacobs, I. M. Tought, E. V. Kuensberg, and J. D. E. Knox. 1966. *Chromosome studies on adults,* vol. 42 of *Eugenics lab memoirs.* London: Cambridge University Press.

Cowan, Ruth Schwartz. 1992. Genetic technology and reproductive choice: An ethics for autonomy. In *The code of codes: Scientific and social issues in the Human Genome Project,* 245–263. Cambridge: Harvard University Press.

Dancis, Joseph. 1973 The prenatal detection of hereditary defects. In *Medical genetics,* ed. Victor McKusick and Robert Claiborne. New York: H. P. Publishing.

———. 1983. Familial dysautonomia. In *Autonomic failure: A textbook of clinical disorders of the autonomic nervous system,* ed. Roger Bannister, 615–639. Oxford: Oxford University Press.

Daston, Lorraine, and Katherine Park. 1998. *Wonders and the order of nature, 1150–1750.* Cambridge, MA: Zone Books, MIT Press.

de Chadarevian, Soraya. 2002. *Designs for life: Molecular biology after World War II.* Cambridge: Cambridge University Press.

de Chadarevian, Soraya, and Harmke Kamminga, eds. 1998. *Molecularizing biology and medicine: New practices and alliances, 1910s–1970s.* Amsterdam: Harwood Academic.

de la Chapelle, Albert. 1961. Constrictions in normal human chromosomes. *Lancet* 26 (Aug.): 460–464.

Dell'Aringa, Stefanie. 1997. Dr. Axelrod brings hope to patients with rare disease. *American Academy of Pediatrics News* 10 (5).

Denver Study Group. 1960a. A proposed standard nomenclature of human mitotic chromosomes. *Lancet* 1:1063–1065.

———. 1960b. A proposed standard system of nomenclature of human mitotic chromosomes (letter to the editor). *American Journal of Human Genetics* 12 (4): 384–388.

Down, J. Langdon. 1995. Observations on an ethnic classification of idiots. *Mental Retardation* 33 (1): 54–56.

Ehrlich, Paul. 1971. *The population bomb*. New York: Ballantine.

Emery, Alan E. H. 1995. *The history of a genetic disease: Duchenne muscular dystrophy or Meryon's disease*. London: Royal Society of Medicine Press.

Epstein, Steven. 1996. *Impure science: AIDS, activism, and the politics of knowledge*. Berkeley and Los Angeles: University of California Press.

Essen-Möller, E. 1970. The twin register of Lund. *Acta Geneticae Medicae et Gemellologiae* 19:355.

Faden, Ruth R., Neil Holtzman, and Judith Chwalow. 1982. Parental rights, child welfare and public health: The case of PKU screening. *American Journal of Public Health* 72 (12): 1396–1400.

Farber, Susan. 1981. *Identical twins reared apart: A reanalysis*. New York: Basic Books.

Ferguson-Smith, M. A. 1962. The identification of human chromosomes. *Proceedings of the Royal Society of Medicine* 55:471–475.

Ferguson-Smith, M. A., Marie E. Ferguson-Smith, Patricia M. Ellis, and Marion Dickson. 1962. The sites and relative frequencies of secondary constrictions in human somatic chromosomes. *Cytogenetics* 1:325–343.

Ferguson-Smith, M. A., A. W. Johnston, and A. N. Weinberg. 1960. Primary amentia and micro-orchidism associated with a XXXY sex chromosome constitution. *Lancet* 2:184–187.

Fisher Family. 1957. *Descendants and history of Christian Fisher family*. Ronks, PA: Amos L. Fisher.

Følling, Asbjørn. 1934. Phenylpyruvic acid as a metabolic anomaly in connection with imbecility. *Nordisk Med Tidskrift* 8:1054–1059.

Ford, C. E., and J. L. Hamerton. 1956. A colchicine, hypotonic citrate squash sequence for mammalian chromosomes. *Stain Technology* 31:247–251.

Ford, C. E., P. A. Jacobs, and L. G. Lajtha. 1958. Human somatic chromosomes. *Nature* 181:1565.

Ford, C. E., K. W. Jones, O. J. Miller, U. Mittwoch, L. S. Penrose, M. Ridler, and A. Shapiro. 1959. The chromosomes in a patient showing both mongolism and Klinefelter syndrome. *Lancet* 1:709–710.

Forster, W., and M. Tyndel. 1956. The neuropsychiatric aspects of familial dysautonomia (Riley-Day syndrome). *Journal of Mental Science* 102:345.

Franklin, Sarah. 2003. Rethinking nature-culture: anthropology and the new genetics. *Anthropological Theory* 3 (1): 65–85.

Franklin, Sarah, and Margaret Lock, eds. 2003. *Remaking life and death: Toward an anthropology of the biosciences*. Santa Fe: School of American Research Press.

Fraumeni, Joseph. 2000. The gene versus environment debate. *Linkage* (newsletter of the Division of Cancer Epidemiology and Genetics, National Cancer Institute, Bethesda) 11 (Dec.): 1–2.

Freeman, A. M., W. Heine, and J. Havel. 1957. Psychological aspects of familial dysautonomia. *American Journal of Orthopsychiatry* 27:96.

Freireich, Emil, and Hagop Kantarjian, eds. 1993. *Leukemia: Advances in research and treatment.* Boston: Kluwer Academic.

Fujimura, Joan H. 1996. *Crafting science: A sociohistory of the quest for the genetics of cancer.* Cambridge: Harvard University Press.

Fuller, John L., and W. Robert Thompson. 1960. *Behavior genetics.* New York: Wiley.

Galton, Francis. 1874. *English men of science: Their nature and nurture.* London: Macmillan. Facsimile edition, Bristol, England: Thoemmes, 1998.

———. 1875. The history of twins as a criterion of the relative powers of nature and nurture. *Fraser's Magazine* 12:566–576.

Garrod, Archibald E. 1909. *Inborn errors of metabolism.* Oxford: Academic Press.

Garwood, D. S., and B. Augenbraun. 1968. Coordinated psychotherapeutic approaches to a familial dysautonomic preschool boy and his parents. *Psychoanalytic Review* 55:62.

Gedda, Luigi, Paolo Parisi, and Walter E. Nance, eds. 1981. *Twin research 3: Proceedings of the Third International Congress on Twin Studies* (3 vols). New York: Alan R. Liss.

Gerhard, D. S., M. C. LaBuda, S. D. Bland, C. Allen, J. A. Egeland, and D. L. Pauls. 1994. Initial report of a genome search for the affective disorder predisposition gene in the old order Amish pedigrees: Chromosomes 1 and 11. *American Journal of Medical Genetics* 54 (4): 398–404.

Giannelli, F., and R. M. Howlett. 1966. The identification of the chromosomes of the D group (13–15) Denver: An autoradiographic and measurement study. *Cytogenetics* 5:186–205.

Gilbert, C. S. 1966. A computer program for the analysis of human chromosomes. *Nature* 212:1437–1440.

Gilbert, Scott. 1997. Bodies of knowledge: Biology and the intercultural university. In *Changing life: Genomes, ecologies, bodies, commodities,* ed. Peter J. Taylor, Saul E. Halfon, and Paul N. Edwards, 36–55. Minneapolis: University of Minnesota Press.

Ginns, E. I., J. Ott, J. A. Egeland, C. R. Allen, C. S. Fann, D. L. Pauls, J. Weissenbachoff, J. P. Carulli, K. M. Falls, T. P. Keith, and S. M. Paul. 1996. A genome-wide search for chromosomal loci linked to bipolar affective disorder in the Old Order Amish. *Nature Genetics* 12 (4): 431–435.

Ginsburg, Faye, and Rayna Rapp. 1999. Fetal reflections: Confessions of two feminist anthropologists as mutual informants. In *Fetal positions / Feminist practices,* ed. Lynn Morgan and Meredith Michaels, 279–295. Philadelphia: University of Pennsylvania Press.

Glass, Bentley. 1988. *A guide to the genetics collections of the American Philosophical Society.* Philadelphia: American Philosophical Society Library.

Goedde, H. W., and D. P. Agarwal, eds. 1987. *Genetics and alcoholism.* New York: Alan R. Liss.

———. 1989. *Alcoholism: Biomedical and genetic aspects*. New York: Pergamon Press.

Goldsmith, H. Hill. 1991. A zygosity questionnaire for young twins: A research note. *Behavior Genetics* 21 (3): 257–269.

Goodman, Richard M. 1979. *Genetic disorders among the Jewish people*. Baltimore: Johns Hopkins University Press.

Goodwin, D. W. 1979. Alcoholism and heredity: A review and hypothesis. *Archives of General Psychiatry* 36:57–61.

Gottesman, Irving I. 1984. Eliot Slater (1904–1983): An appreciation. *Behavior Genetics* 14 (2): 107–110.

Gottesman, I., and A. Bertelsen. 1996. The legacy of German psychiatric genetics: Hindsight is always 20/20. *American Journal of Medical Genetics (Neuropsychiatric Genetics)* 67:317–322.

Gottesman, Irving I., and J. Shields. 1972. *Schizophrenia and genetics: A twin study vantage point*. New York: Academic Press.

Gottesman, Michael. 2003. Cancer gene therapy: An awkward adolescence. *Cancer Gene Therapy* 10 (7): 501–508.

Gottweiss, Herbert. 1998. *Governing molecules: The discursive politics of genetic engineering in Europe and the United States*. Cambridge, MA: MIT Press.

Guthrie, Robert. 1961. Blood screening for phenylketonuria (letter to the editor). *Journal of the American Medical Association* 178:863.

Guthrie, Robert, and Ada Susi. 1963. A simple phenylalanine method for detecting phenylketonuria in large populations of newborn infants. *Pediatrics* 32:338–343.

Guthrie, Robert, and H. Tieckelmann. 1962. The inhibition assay: Its use in screening urinary specimens for metabolic differences associated with mental retardation. In *Proceedings of the London Conference on the Scientific Study of Mental Deficiency*, 672–677. London: May and Baker.

Guthrie, Robert, and Stewart Whitney. 1964. *Phenylketonuria: Detection in the newborn infant as a routine hospital procedure*. Washington, DC: Children's Bureau, U.S. Department of Health, Education and Welfare.

Hamerton, John. L. 1971. *Human cytogenetics*. New York: Academic Press.

Hamlin, Christopher, and Philip T. Shepard. 1993. *Deep disagreement in U.S. agriculture: Making sense of policy conflict*. Boulder, CO: Westview.

Haraway, Donna. 2003. For the love of a good dog: Webs of action in the world of dog genetics. In *Genetic nature/culture: Anthropology and science beyond the two culture divide*, ed. Alan Goodman, Deborah Heath, and Susan Lindee, 111–131. Berkeley and Los Angeles: University of California Press.

Harris, Harry. 1959. *Human biochemical genetics*. Cambridge: Cambridge University Press.

———. 1975. *Prenatal diagnosis and selective abortion*. Cambridge: Harvard University Press.

Hauge M., B. Harvald, M. Fischer, K. Gotlieb-Jensen, N. Juel-Nielsen, I. Raebild, R. Shapiro, and T. Videbech T. 1968. The Danish twin register. *Acta Geneticae Medicae et Gemellologiae* 17 (2): 315–332.

Hayes, Richard. 2002. The market for modified humans. *World Watch* 15 (4): 11–12.

Heath, Andrew C., Pamela A. F. Madden, and Kathleen K. Bucholz. 1999. Ascertainment of a twin sample by computerized record matching, with assessment of possible sampling biases. *Behavior Genetics* 29 (4): 209–219.

Henkin, R. I., and I. J. Kopin. 1964. Abnormalities of taste and smell thresholds in familial dysautonomia: Improvement with methacholine. *Life Sciences* 3:1319–1325.

Hessenbruch, Arne. 1999. The spread of precision measurement in Scandinavia, 1660–1800. In *Sciences in the European periphery during the enlightenment*, ed. Kostas Gavroglu, 179–224. Dordrecht, Netherlands: Kluwer.

Hilgartner, Stephen. 1998. Data access policy in genome research. In *Private science: Biotechnology and the rise of the molecular sciences*, ed. Arnold Thackray. Philadelphia: University of Pennsylvania Press.

Hisada, Michie, Judy Garber, Clair Fung, Joseph Fraumeni, and Frederick Li. 1998. Multiple primary cancers in families with Li-Fraumeni syndrome. *Journal of the National Cancer Institute* 90 (8): 606–611.

Horner, F., and C. Streamer. 1956. Effects of phenylalanine-restricted diet on patients with phenylketonuria: Clinical observations in three cases. *Journal of the American Medical Association* 161:1628–1630.

Hostetler, J. A. 1963. *Amish society*. Baltimore: Johns Hopkins University Press.

———. 1963–1964. Folk and scientific medicine in Amish society. *Human Organization* 22 (4): 269–275.

Hrubec, Zdenek, and James V. Neel. 1978. The National Academy of Sciences–National Research Council Twin Registry: Ten years of operation. In *Twin research: Proceedings of the Second International Congress on Twin Studies, August 29–September 1, 1977, Washington, D.C.*, ed. Walter E. Nance; assoc. eds Gordon Allen and Paolo Parisi, 153–172. New York: Alan R. Liss.

Hsia, David Yi-Yung. 1960. Recent developments in inborn errors of metabolism. *Journal of Public Health* 50 (10): 1653–1661.

Hsu, T. C. 1952. Mammalian chromosomes in vitro: I. The karyotype of man. *Journal of Heredity* 43:167–172.

———. 1979. *Human and mammalian cytogenetics: An historical perspective*. Berlin: Springer-Verlag.

Hungerford, David. 1961. A study of the chromosomes in leukocytes from the peripheral blood of children with leukemia (Ph.D. diss., University of Pennsylvania).

Hungerford, D. A., S. Makino, M. Sasaki, A. A. Awa, and Gloria Balaban. 1969. Chromosome studies of the Ainu population of Hokkaido. *Cytogenetics* 8:74–79.

Hunt, Bruce J. 1994. The Ohm is where the art is: British telegraph engineers and the development of electrical standards. *Osiris* 9:48–63.

Jablon, Seymour, James V. Neel, H. Gershowitz, and G. F. Atkinson. 1967. The NAS-NRC Twin Panel: Methods of construction of the panel, zygosity diagnosis and proposed uses. *American Journal of Human Genetics* 19 (Mar.): 133–161.

Jacobs, Paul. 2000. A father's mission. *San Jose Mercury News*, 31 Dec.

Jacobs, P. A., A. G. Baikia, W. M. Court-Brown, D. M. MacGregor, M. MacLean, and D. G. Harnden. 1959. Evidence for the existence of a human "super female." *Lancet* 2:423–425.

Jacobs, P. A., D. G. Harnden, W. M. Court Brown, J. Goldstein, H. G. Close, T. N. Macgregor, N. Maclean, and J. A. Strong. 1960. Abnormalities involving the X-chromosome in women. *Lancet* 1:1213–1216.

Jensen, Arthur. 1969. How much can we boost IQ and scholastic achievement? *Harvard Educational Review* 19 (Feb.): 1–123.

———. 1974. Kinship correlations reported by Sir Cyril Burt. *Behavior Genetics* 4 (1): 1–28.

Jervis, George. 1966. Overall view of our biochemical knowledge of PKU. In *Phenylketonuria and allied metabolic diseases*, ed. John A. Anderson and Kenneth F. Swaiman, 33. Washington, DC: U.S. Children's Bureau.

Kaback, Michael. 1977. *Tay Sachs disease: Screening and prevention.* New York: John Wiley and Sons.

Kaij, Lennart. 1960. *Alcoholism in twins.* Stockholm: Almqvist and Wiksell International.

Kallmann, Franz J. 1938. *The genetics of schizophrenia: A study of heredity and reproduction in the families of 1,087 schizophrenics,* with the assistance of Senta Jonas Rypins and introduction by Nolan D. C. Lewis. New York: J. J. Augustin.

———. 1946. The genetic theory of schizophrenia: An analysis of 691 schizophrenic twin index families. *American Journal of Psychiatry* 103:309–322.

———. 1950a. The genetics of psychoses: An analysis of 1,232 twin index families. *American Journal of Human Genetics* 2:385–390.

———. 1950b. The genetics of psychoses: An analysis of 1,232 twin index families. In *Congres internationale de psychiatrie,* 1–27. Paris: Herman.

———. 1952. Human genetics as a science, as a profession, and as a social-minded trend of orientation. *American Journal of Human Genetics* 4 (4): 237–245.

———. 1953. *Heredity in health and mental disorder: Principles of psychiatric genetics in the light of comparative twin studies,* with foreword by Nolan D. C. Lewis. New York: W. W. Norton.

Kallmann, F. J., J. Deporte, E. Deporte, and L. Feingold. 1949. Suicide in twins and only children. *American Journal of Human Genetics* 1:113–126.

Kay, Lily E. 1993. *The molecular vision of life: Caltech, the Rockefeller Foundation and the rise of the new biology.* New York: Oxford University Press.

———. 2000. *Who wrote the book of life? A history of the genetic code.* Stanford: Stanford University Press.

Keller, Evelyn Fox. 1995. *Refiguring life: Metaphors of 20th-century biology.* New York: Columbia University Press.

———. 2000. *The century of the gene.* London: Harvard University Press.

Kevles, Daniel J. 1985. *In the name of eugenics: Genetics and the uses of human heredity.* New York: Alfred A. Knopf.

Kevles, Daniel J., and Leroy Hood, eds. 1992. *The code of codes: Scientific and social issues in the Human Genome Project.* Cambridge: Harvard University Press.

Klinger, Harold P., and Orlando J. Miller. 1968. Prenatal sex chromatin and chromosome analysis. In *Diagnosis and treatment of fetal disorders,* ed. Karlis Adamsons, 72–82. New York: Springer-Verlag.

Koch, Jean Holt. 1997. *Robert Guthrie: The PKU story.* Pasadena, CA: Hope Publishing.

Kohler, Robert. 1994. *Lords of the fly: Drosophila genetics and the experimental life.* Chicago: University of Chicago Press.

Kolata, Gina. 1996. Parents take charge, putting gene hunt onto the fast track. *New York Times,* 16 July.

Kottler, Malcolm J. 1974. From 48 to 46: Cytological technique, preconception and the counting of human chromosomes. *Bulletin of the History of Medicine* 48:467–471.

Kritchman, Marilyn M., Herman Schwartz, and Emanuel M. Papper. 1959. Experiences with general anesthesia in patients with familial dysautonomia. *Journal of the American Medical Association* 170 (5): 529–533.

Kuklick, Henrika, and Robert Kohler. 1996. Introduction. In *Science in the field,* vol. 11 of *Osiris,* 2nd ser., ed. Henrika Kuklick and Robert Kohler, 1–14. Chicago: University of Chicago Press.

LaBuda, M. C., M. Maldonado, D. Marshall, K. Otten, and D. S. Gerhard. 1996. A follow-up report of a genome search for affective disorder predisposition loci in the Old Order Amish. *American Journal of Human Genetics* 59 (6): 1343–1362.

LaDu, Bert N., R. Rodney Howell, Patricia J. Michael, and Eva K. Sober. 1963. A quantitative micromethod for the determination of phenylalanine in blood and its application to the diagnosis of phenylketonuria in infants. *Pediatrics* 31 (1): 39–57.

Lawrence, M. M. 1956. Comprehensive approach to study of a case of familial dysautonomia. *Psychoanalytic Review* 43:358.

Lejeune, J., M. Gautier, and R. Turpin. 1959a. Les chromosomes humane en culture de tissus. *Comptes Rendus de l'Academie de Sciences* 248:602–603.

———. 1959b. The chromosomes of man. *Lancet* 1:885.

———. 1959c. Etude des chromosomes somatiques de neuf enfants mongoliens. *Comptes Rendus de l'Academie de Sciences* 248:1721–1722.

Lejeune, Jerome, and Raymond Turpin. 1961. Chromosomal aberrations in man (originally presented as part of a symposium on genetics held at Western Reserve University, 10–12 Oct., 1960). *American Journal of Human Genetics* 13 (Mar.): 175–183.

Lejeune, Jerome, Raymond Turpin, and Marthe Gautier. 1959. Mongolisme: une maladie chromosomique (trisomy). *Bulletin de l'Academie Nationale de Medecine* 143:256–265.

Lennox, Bernard 1961. Chromosomes for beginners. *Lancet,* 31 May, 1046–1051.

Le Saux, O., K. Beck, C. Sachsinger, C. Silvestri, C. Treiber, H. H. Goring, E. W. Johnson, A. De Paepe, F. M. Pope, I. Pasquali-Ronchetti, L. Bercovitch, A. S. Marais, D. L. Viljoen, S. F. Terry, and C. D. Boyd. 2001. A spectrum of ABCC6 mutations is responsible for pseudoxanthoma elasticum. *American Journal of Human Genetics* 69 (4): 749–764.

Levan, Albert, and T. C. Hsu. 1959. The human idiogram. *Hereditas* 45:665–674.

Lewis, Margaret Reed. 1932. Reversible solution of the mitotic spindle of living chick embryo cells studied in vitro. *Archives für Experimentelle Zellforschung* 16:159–166.

Li, Frederick, and Joseph Fraumeni. 1969. Soft-tissue sarcomas, breast cancer, and other neoplasms: A familial syndrome? *Annals of Internal Medicine* 71 (4): 747–752.

Li, J. G., and E. E. Osgood. 1949. A method for the rapid separation of leukocytes and nucleated erthrocytes from blood or marrow with a phytohemagglutinin from red beans (*Phaseolus vulgaris*). *Blood* 4:670–675.

Lichtenstein, P., H. V. Holm, P. K. Verkasalo, et al. 2000. Environmental and heritable factors in the causation of cancer—analyses of cohorts of twins from Sweden, Denmark, and Finland. *New England Journal of Medicine* 343:78–85.

Liebler, Daniel C. 2002. *Proteomics: An introduction to the new biology.* Totowa, NJ: Humana.

Liebman, Sumner D. 1957. Riley-Day syndrome (familial dysautonomia): Concerning the etiology of the corneal pathology; an ocular survey of nineteen cases. *Archives of Ophthalmology* 58 (6): 188–192.

Lindee, M. Susan. 1994. *Suffering made real: American science and the survivors at Hiroshima.* Chicago: University of Chicago Press.

———. 2002. Genetic disease in the 1960s: A structural revolution. *American Journal of Medical Genetics* 115 (2): 75–82.

———. 2003. Voices of the dead: James Neel's Amerindian studies. In *Lost paradises and the ethics of research and publication,* ed. Francisco Salzano and Magdalene Hurtado, 40–73. New York: Oxford University Press.

Loehlin, John C., and R. C. Nichols. 1976. *Heredity, environment and personality.* Austin: University of Texas Press.

London Conference on the Normal Human Karyotype. 1963. *Cytogenetics* 2:264–268.

Luxenburger, H. 1930. Psychiatrische-Neurologische Zwillingspathologie. *Zentral-blatt für die Gesamte: Neurologische Psychiatrie* 56:145.

Mabry, C. C., J. C. Denniston, T. L. Nelson, and C. D. Son. 1963. Maternal phenylke-tonuria: A cause of mental retardation in children without the metabolic dis-order. *New England Journal of Medicine* 275:1331–1336.

Mackta, Jayne. 1992. *Integrating consumers into the regional genetics networks.* Chevy Chase, MD: Alliance of Genetic Support Groups.

Marks, Jonathan. 2002. *What it means to be 98 percent chimpanzee: Apes, people, and their genes.* Berkeley: University of California Press.

Marshall, E. Jane, and Robin M. Murray. 1989. The contribution of twin studies to alcoholism research. In *Alcoholism: Biomedical and genetic aspects*, ed. H. W. Goedde and D. P. Agarwal, 277–289. New York: Pergamon Press.

Martin, Aryn. 2004. Can't anybody count? Counting as an epistemic theme in the history of human chromosomes. *Social Studies of Science* 34 (4): 1–26.

Mazumdar, Pauline H. 1996. Two models for human genetics: Blood grouping and psychiatry in Germany between the World Wars. *Bulletin of the History of Med-icine* 70:609–657.

McKusick, Victor A. 1949. Broedel's ulnar palsy, with unpublished Broedel sketches. *Bulletin of the History of Medicine* 23:469–479.

———. 1954. The cardiovascular and genetic aspects of Marfan's syndrome, a heri-table disorder of connective tissue. *Bulletin of the Johns Hopkins Hospital* 94:159–161.

———. 1956. *Heritable disorders of connective tissue.* St. Louis: Mosby.

———. 1958. *Cardiovascular sound in health and disease.* Baltimore: Williams and Wilkins.

———. 1962 On the X chromosome of man. *Quarterly Review of Biology* 37:69–175.

———. 1964. *On the X chromosome of man.* Washington, DC: American Institute of Biological Science.

———. 1966 and various years. *Mendelian inheritance in man.* Baltimore: Johns Hop-kins University Press. www.ncbi.nlm.nih.gov/entrez/query.fcgi?db=omim.

———. 1969. On lumpers and splitters, or the nosology of genetic disease. *Per-spectives in Biology and Medicine* 12 (2): 298–312.

———. 1973. *Medical genetics.* New York: H. P. Publishing.

———. 1978. *Medical genetic studies of the Amish: Selected papers assembled with com-mentary.* Baltimore: Johns Hopkins University Press.

———. 1981 The last twenty years: An overview of advances in medical genetics. In *Mammalian genetics and cancer: The Jackson Laboratory Fiftieth Anniversary Symposium*, ed. Elizabeth S. Russell, 127–144. New York: Alan R. Liss.

———. 1992. Human genetics: The last 35 years, the present and the future. *Ameri-can Journal of Human Genetics* 50:663–670.

———. 1993. *Heritable disorders of connective tissue*, 5th ed. St. Louis: Mosby.

————. 1996. History of medical genetics. In *Principles and practices of human genetics*, ed. D. L. Rimoin, J. M. Connor, and R. E. Pyeritz, xiii–xviii. Edinburgh: Churchill Livingston.

————. 2000. Ellis–van Creveld syndrome and the Amish. *Nature Genetics* 24:203–204.

McKusick, V. A., R. Eldridge, J. A. Hostetler, U. Ruangwit, and J. A. Egeland. 1964a. Dwarfism in the Amish: I. The Ellis van Creveld syndrome. *Bulletin of the Johns Hopkins Hospital* 115:306–336.

————. 1964b Dwarfism in the Amish: II. Cartilage-hair hypoplasia. *Bulletin of the Johns Hopkins Hospital* 116:285–326.

McKusick, V. A., J. A. Hostetler, and Janice A. Egeland 1964. Genetic studies of the Amish: Background and potentialities. *Bulletin of the Johns Hopkins Hospital* 115:203–222.

McKusick, V. A., J. A. Hostetler, J. A. Egeland, and R. Eldridge. 1964. The distribution of certain genes in the Old Order Amish. *Cold Spring Harbor Symposia on Quantitative Biology* 29:99–113.

Medin, Douglas L., and Scott Atran. 1999. *Folkbiology.* Cambridge, MA: MIT Press.

Mendelsohn, Everett, Judith Swazey, and Irene Taviss, eds. 1971. *Human aspects of biomedical innovation.* Cambridge: Harvard University Press.

Menees, T. O., J. D. Miller, and L. E. Holly. 1930. Amniography: Preliminary report. *American Journal of Roentgenology and Radiation Therapy* 24:363–366.

Merriman, Curtis. 1924. *The intellectual resemblance of twins.* Princeton: Psychological Review Co.

Miller, Fiona. 1999. A blueprint for defining health: Making medical genetics in Canada, 1930s–1970s (Ph.D. Diss., York University, Toronto).

Milunsky, Aubrey. 1973. *The prenatal diagnosis of hereditary disorders.* Springfield, IL: Charles C Thomas.

————, ed. 1979 and various years. *Genetic disorders and the fetus: Diagnosis, prevention and treatment.* New York: Plenum Press. Second ed., 1986; 3rd, 4th, 5th eds, Baltimore: Johns Hopkins University Press, 1992, 1998, 2004.

Moloshek, R. E., and J. E. Moseley. 1956. Familial dysautonomia: Pulmonary manifestations. *Pediatrics* 17:327.

Moorhead, P. S., P. C. Nowell, W. J. Mellman, D. M. Battips, and D. A. Hungerford. 1960. Chromosome preparation of leukocytes cultured from human peripheral blood. *Experimental Cell Research* 20:613–616.

Morange, Michel. 1998. *A history of molecular biology,* trans. Matthew Cobb and Michel Morange. Cambridge: Harvard University Press.

Moses, S. W., Y. Rotem, N. Jagoda, N. Talmor, F. Eichhorn, and S. Levin. 1967. A clinical, genetic and biochemical study of familial dysautonomia in Israel. *Israel Journal of Medical Science* 3:358.

Motulsky, Arno. 1971. The William Allen Memorial Award Lecture: Human and medical genetics: A scientific discipline and an expanding horizon. *American Journal of Human Genetics* 23 (2): 107–123.

Mourant, A. E., Ada C. Kopec, and Kazimiera Domaniewska-Sobczak. 1978. *The genetics of the Jews.* New York: Clarendon Press.

Muench, Karl H. 1988. *Genetic medicine.* New York: Elsevier.

Muldal, S., and Ch. H. Ockey. 1961. The Denver classification and group II. *Lancet* 2:462–463.

Muller, H. J. 1925. Mental traits and heredity. *Journal of Heredity* 16:433–448.

———. 1949. Progress and prospects in human genetics. *American Journal of Human Genetics* 1 (Sept.): 1–18.

Muller-Hill, Benno. 1998. *Murderous science: Elimination by scientific selection of Jews, gypsies and others in Germany, 1933–1945.* Cold Spring Harbor, NY: Cold Spring Harbor Laboratory Press.

Nance, Walter E., ed. 1978. *Twin research: Proceedings of the Second International Congress on Twin Studies, August 29–September 1, 1977, Washington, D.C.,* assoc. eds Gordon Allen and Paolo Parisi. New York: Alan R. Liss.

Nathan, David G. 1995. *Genes, blood and courage: A boy called Immortal Sword.* Cambridge: Belknap Press of Harvard University Press.

National Foundation for Jewish Genetic Diseases. 1980. *You have a right to know . . . about Jewish genetic disease.* New York: The Foundation.

National Panel of Consultants on the Conquest of Cancer. 1971. *National Program for the Conquest of Cancer: Report to the Committee on Labor and Public Welfare, U.S. Senate.* Washington, DC: Government Printing Office.

National Tay-Sachs and Allied Diseases Association. 1994. *The home care book: A parent's guide to caring for children with progressive neurological diseases.* Brookline, MA: The Association.

Neel, James V. 1949. The inheritance of sickle cell anemia. *Science* 110:64–66.

———. 1994. *Physician to the gene pool.* New York: Wiley.

Neel, J. V., M. Shaw, and W. J. Schull. 1965. *Genetics and epidemiology of chronic diseases* (Public Health Service Bulletin No. 1163). Washington, DC: US Public Health Service.

Nelkin, Dorothy, and M. Susan Lindee. 1995. *The DNA mystique: The gene as a cultural icon.* New York: W. H. Freeman.

Neurath, P. W., ed. 1967. *I-CAN International—Chromosome Analysis Newsletter,* no. 105. Vienna: International Atomic Energy Agency.

Neurath, P. W., and K. Enslein. 1969. Human chromosome analysis as computed from arm lengths measurement. *Cytogenetics* 8:337–354.

Newman, Horatio, Frank Freeman, and Karl Holzinger. 1937. *Twins.* Chicago: University of Chicago Press.

New York State Department of Health. 1980. The Dysautonomia Treatment and Evaluation Center at NYU Medical Center. *Genesis* (newsletter of the Genetics Network of the Empire State) 11:2.

Nora, James J., F. Clarke Fraser, John Bear, Cheryl R. Greenberg, David Patterson, and Dorothy Warburton. 1994. *Medical genetics: Principles and practices*, 4th ed. Philadelphia: Lea and Febiger.

Nowell, Peter. 1998. Cancer genetics, cytogenetics: Defining the enemy within (personal essay). *Nature Medicine* 4 (10): 1107–1108.

Nukaga, Yoshio, and Alberto Cambrosio. 1997. Medical pedigrees and the visual production of family disease in Canadian and Japanese genetic counseling practice. In *The sociology of medical science and technology*, ed. Mary Ann Elston, 29–55. Oxford: Blackwell.

Page, W. F. 1995. Annotation: The National Academy of Sciences–National Research Council Twin Registry. *American Journal of Public Health* 85:617–618.

Painter, T. S. 1921. The Y chromosome in mammals. *Science* 53:503–504.

———. 1923. Studies in mammalian spermatogenesis: II. The spermatogenesis of man. *Journal of Experimental Zoology* 37:291–334.

Paris Conference. 1971. Standardization in human cytogenetics. *Birth Defects Original Article Series* 8:7. Reprinted in *Cytogenetics* 11 (1972): 313–362.

Parisi, P., and M. Di Bacco. 1968. Fingerprints and the diagnosis of zygosity in twins. *Acta Geneticae Medicae et Gemellologiae* 17 (2): 333–358.

Partanen, J., K. Bruun, and T. Markkanen. 1966. *Inheritance of drinking behavior*. Helsinki: Finnish Foundation for Alcohol Studies.

Patau, Klaus. 1960. The identification of individual chromosomes, especially in man. *American Journal of Human Genetics* 12:250–276.

———. 1961. Chromosome identification and the Denver report. *Lancet* 1:933–934.

———. 1963. Review of John L. Hamerton, ed. Chromosomes in Man. *Cytogenetics* 2:269–270.

———. 1965. The chromosomes. *Birth Defects Original Article Series* 1 (2): 71–74.

Paul, Diane. 1991. The Rockefeller Foundation and the origins of behavior genetics. In *The expansion of American biology*, ed. Keith R. Benson, Jane Maienschein, and Ronald Rainger, 262–283. New Brunswick, NJ: Rutgers University Press.

———. 1995. *Controlling human heredity, 1865 to the present*. Atlantic Highlands, NJ: Humanities Press.

———. 1998. *The politics of human heredity: Essays on eugenics, biomedicine and the nature-nurture debate*. Albany: State University of New York Press.

———. 1999a. Contesting consent: the challenge to compulsory neonatal screening for PKU. *Perspectives in Biology and Medicine* 24 (2): 207–219.

———. 1999b. Report on the history of newborn screening. In *Promoting safe and effective genetic testing in the United States: Final report*, Task Force on Genetic

Testing, ed. Neil A. Holtzman and Michael S. Watson. Baltimore: Johns Hopkins University Press. (Orig. pub. Washington DC: National Institutes of Health–Department of Energy Working Group on Ethical, Legal and Social Implications of Human Genome Research, 1997.)

Paul, Diane, and Paul J. Edelson. 1998. The struggle over metabolic screening. In *Molecularizing biology and medicine: New practices and alliances, 1910s–1970s*, ed. Soraya de Chadarevian and Harmke Kamminga, 203–220. Amsterdam: Harwood Academic.

Pearson, J., L. Brandeis, and A. C. Cuello. 1982. Depletion of substance P-containing axons in substantia gelatinosa of patients with diminished pain sensitivity. *Nature* 295:61–63.

Pearson, J., G. Budzilovich, and M. J. Finegold. 1971. Sensory, motor, and autonomic dysfunction: The nervous system in familial dysautonomia. *Neurology* 21:486–493.

Peele, Stanton. 1986. The implications and limitations of genetic models of alcoholism and other addictions. *Journal of Studies on Alcohol* 47:63–73.

Peeters, Hilde, Sophie van Gestel, Robert Vlietinck, Catherine Derom, and R. Derom. 1998. Validation of a telephone zygosity questionnaire in twins of known zygosity. *Behavior Genetics* 28 (3): 159–163.

Pemberton, Stephen. 2001. Normality within limits: Hemophilia, the citizen-patient and the risks of medical management in the United States of America from World War II to the age of AIDS (Ph.D. diss., University of North Carolina, Chapel Hill).

Penrose, Lionel. 1934. *The influence of heredity on disease*. London: H. K. Lewis.

———. 1946. Phenylketonuria: A problem in eugenics. *Lancet,* June 29.

———. 1964. A note on the mean measurements of human chromosomes. *Annals of Human Genetics* 28:195–196.

Perlman, Max, Sam Benady, and Ephraim Saggi. 1979. Neonatal diagnosis of familial dysautonomia. *Pediatrics* 63 (2): 238–241.

Plomin, R., M. J. Owen, and P. McGuffin. 1994. The genetic basis of complex human behaviors. *Science* 264 (5166): 1733–1739.

Pollen, Daniel E. 1993. *Hannah's heirs: The quest for the genetic origins of Alzheimer's disease*. New York: Oxford University Press.

Porter, Ian H. 1974. Introduction. In *Clinical cytogenetics and genetics: Boston conference on the diagnosis, genetics and management of birth defects*, ed. Daniel Bergsma, x. New York: Stratton Intercontinental Medical Book.

Porter, Theodore. 1996. *Trust in numbers: The pursuit of objectivity in science and public life*. Princeton: Princeton University Press.

President's Cancer Panel. 1990. *President's Cancer Panel: National Cancer Program Human gene therapy meeting*. Silver Spring, MD: Eberlin Reporting Service.

Proctor, Robert. 1989. *Racial hygiene: Medicine under the Nazis.* Cambridge: Harvard University Press.

———. 1995. *Cancer wars: How politics shapes what we know and don't know about cancer.* New York: Basic Books.

Pyeritz, Reed. 1998. Medical genetics: End of the beginning or beginning of the end? *Genetics in Medicine* 1 (1): 56–60.

Rabinow, Paul. 1996. *Making PCR: A history of biotechnology.* Chicago: University of Chicago Press.

———. 1999. *French DNA: Trouble in purgatory.* Chicago: University of Chicago Press.

Rader, Karen. 2004. *Making mice: Standardizing animals for American biomedical research, 1900–1955.* Princeton: Princeton University Press.

Rapp, Rayna. 2000. *Testing women, testing the fetus: The social impact of amniocentesis in America.* New York: Routledge.

———. 2003. Cell life and death, child life and death: Genomic horizons, genetic diseases, family stories. In *Remaking life and death: Toward an anthropology of the biosciences,* ed. Sarah Franklin and Margaret Lock, 129–164. Santa Fe: School of American Research Press.

Rende, R. D., R. Plomin, and S. G. Vandenberg. 1990. Who discovered the twin method? *Behavior Genetics* 2 (Mar. 20): 277–285.

Riis, P., and F. Fuchs. 1960. Antenatal determination of fetal sex in prevention of hereditary diseases. *Lancet* 2:180–182.

Riley, Conrad. 1956. *Living with a child with familial dysautonomia.* New York: Dysautonomia Foundation.

———. 1964. Maturation of a clinical entity. *New England Journal of Medicine* 9 (27 Aug.): 271. (Unattributed but written by Riley, personal communication.)

———. 1970. A short history of familial dysautonomia (unpublished report to the Dysautonomia Association, 19 Nov.).

Riley, C. M., R. L. Day, D. M. Greeley, and W. S. Langford. 1949. Central autonomic dysfunction with defective lacrimation: Report of five cases. *Pediatrics* 3:468.

Riley, C. M., A. M. Freedman, and W. S. Langford. 1954. Further observations on familial dysautonomia. *Pediatrics* 14:475.

Riley, C. M., and R. H. Moore. 1966. Familial dysautonomia differentiated from related disorder: Case reports and discussions of current concepts. *Pediatrics* 37:43.

Ritvo, Harriet. 1987. *The animal estate: The English and other creatures in the Victorian age.* Cambridge: Harvard University Press.

Rosanoff, A. J., L. M. Handy, and I. A. Rosanoff. 1934. Etiology of epilepsy with special reference to its occurrence in twins. *Archives of Neurology and Psychiatry* 31:1165–1193.

Rosenberg, Charles E. 1997. The bitter fruit: Heredity, disease and social thought.

In *No other gods: On science and American social thought.* Baltimore: Johns Hopkins University Press. (Orig. pub. 1976.)

Rothman, Barbara Katz. 1986. *The tentative pregnancy: Prenatal diagnosis and the future of motherhood.* New York: Viking.

Rowe-Richmond, LaVelda, and LaVona Rowe-Richmond. 1976. The history of the International Twins Association. *Acta Geneticae Medicae et Gemellologiae* 25:387–388.

Rowley, Janet. 1973. A new consistent chromosome abnormality in chronic myelogenous leukemia. *Nature* 243:290–293.

———. 1998. Cancer genetics, cytogenetics: Defining the enemy within (personal essay). *Nature Medicine* 4 (10): 1107–1108.

Rubin, Berish. 2003. Tocotrienols induce IKBKAP expression: A possible therapy for familial dysautonomia. *Biochemical and Biophysical Research Communications* 306 (1): 303–309.

Sak, H. G., A. A. Smith, and J. Dancis. 1967. Psychometric evaluations of children with familial dysautonomia. *American Journal of Psychiatry* 124:5.

Salzano, Francisco. 1957. The blood groups of South American Indians. *American Journal of Physical Anthropology* 15:555–579.

Sandberg, Avery A. 1979. Before 1956: Some historical background to the study of chromosomes in human cancer and leukemia. *Cancer Genetics and Cytogenetics* 1:87–94.

Scarr, Sandra. 1987. Three cheers for behavior genetics: Winning the war and losing our identity. *Behavior Genetics* 17 (3): 219–228.

Schaffer, Simon. 1994. Rayleigh and the establishment of electrical standards. *European Journal of Physics* 15:277–285.

———. 1995. Accurate measurement is an English science. In *The values of precision,* ed. Norton Wise, 135–172. Princeton: Princeton University Press.

Schull, William J. 1990. *Song among the ruins.* Cambridge: Harvard University Press.

Scriver, Charles R. 1995. What ever happened to PKU? *Clinical Biochemistry* 28 (2): 137–144.

Secord, James. 1981. Nature's fancy: Charles Darwin and the breeding of pigeons. *Isis* 72:162–186.

Seixas, F. A., G. S. Omenn, E. D. Burk, and S. Eggleston, eds. 1972. *Nature and Nurture in Alcoholism,* vol. 197 of *Annals of the New York Academy of Sciences.* New York: New York Academy of Sciences.

Sexton, Sarah. 2002. A deceptive promise of cures for disease. *World Watch* 15 (4): 18–21.

Shapiro, Burton L., and Ralph C. Heusner. 1991. *A parent's guide to cystic fibrosis.* Minneapolis: University of Minnesota Press.

Shettles, L. B. 1956. Nuclear morphology of cells in human amniotic fluid in relation to the sex of the infant. *American Journal of Obstetrics and Gynecology* 71:834–838.

Siemens, Hermann W. 1924. *Die Zwillingspathologie*. Berlin: Springer.

Slater, Eliot. 1953. *Psychotic and neurotic illnesses in twins*, with the assistance of James Shields (Medical Research Council Special Report Series no. 278). London: Stationery Office.

Slaugenhaupt, Susan A., Anat Blumenfeld, Sandra P. Gill, Maire Leyne, James Mull, Math P. Cuajungco, Christopher B. Liebert, Brian Chadwick, Maria Idelson, Luba Reznik, Christiane M. Robbins, Izabele Makalowska, Michael Brownstein, Daniel Krappmann, Claus Scheidereit, Channa Maayan, Felicia B. Axelrod, and James F. Gusella. 2001. Tissue-specific expression of a splicing mutation in the IKBKAP gene causes familial dysautonomia. *American Journal of Human Genetics* 68:598–605.

Slifer, Eleanor. 1934. Insect development: VI. The behavior of grasshopper embryos in anisotonic balanced salt solutions. *Journal of Experimental Zoology* 67:137–157.

Smith, Alfred A., and Joseph Dancis. 1962. Response to intradermal histamine in familial dysautonomia: A diagnostic test. *Pediatrics* 63 (5): 889–894.

———. 1964a. Peripheral sensory deficits in familial dysautonomia. *Journal of Pediatrics* 65:1035.

———. 1964b. Familial pheochromocytoma presenting as familial dysautonomia. *Pediatrics* 65 (3): 463–465.

———. 1970. Familial dysautonomia: What's in a name? *Pediatrics* 77 (1): 174–175.

Smocovitis, Vassiliki. 1996. *Unifying biology: The evolutionary synthesis and evolutionary biology*. Princeton: Princeton University Press.

Snow, Herbert. 1885. Is cancer hereditary? *British Medical Journal*, 10 Oct., 690–692.

Spuhler, J. N., ed. 1967. *Genetic diversity and human behavior* (Viking Fund Publications in Anthropology no. 45). New York: Wenner-Gren Foundation for Anthropological Research.

Stern, Curt. 1949. *Principles of human genetics*. San Francisco: W. H. Freeman.

Stranahan, Susan Q. 1997. Clinic a lifeline to children: A doctor finds his calling among the Amish and Mennonites. *Philadelphia Inquirer*, 15 Sept.

Sumner, A. T. 1982 The nature and mechanisms of chromosome banding. *Cancer Genetics and Cytogenetics* 6:59–87.

Task Force on Genetic Testing. 1999. *Promoting safe and effective genetic testing in the United States: Final report*, ed. Neil A. Holtzman and Michael S. Watson. Baltimore: Johns Hopkins University Press. (Orig. pub. Washington DC: National Institutes of Health–Department of Energy Working Group on Ethical, Legal and Social Implications of Human Genome Research, 1997.)

Tate, Carolyn, and Gordon Bendersky. 1999. Olmec sculptures of the human fetus. *Perspectives in Biology and Medicine* 42 (3): 303–332.

Tauber, Miriam. 2001. Race to discover gene mutation ends in virtual tie. *Forward*, 17 Aug.

Taubman, Paul. 1976a. The determinants of earnings: Genetics, family and other environments. *American Economic Review* 66:858–870.

―――. 1976b. Earnings, education, genetics and environment. *Journal of Human Resources* 11:447–461.

―――. 1978. Determinants of socioeconomic success: Regression and latent variables analysis in samples of twins. In *Twin research: Proceedings of the Second International Congress on Twin Studies, August 29–September 1, 1977, Washington, D.C.,* ed. Walter E. Nance; assoc. eds Gordon Allen and Paolo Parisi, 175–187. New York: Alan R. Liss.

Taussig, Karen-Sue. 2005. Molecules, medicine and bodies: Building social relationships for a molecular revolution in medicine. In *Complexities: Beyond nature and nurture,* ed. Susan McKinnon and Sydel Silverman. Chicago: University of Chicago Press.

Taussig, Karen-Sue, Rayna Rapp, and Deborah Heath. 2002. Flexible eugenics: Technologies of the self in the age of genetics. In *Anthropology in the age of genetics: Practice, discourse, critique,* ed. A. Goodman, D. Heath, S. Lindee, 58–76. Berkeley and Los Angeles: University of California Press.

Taylor, Charlotte. 1970. Marriages of twins to twins. *Acta Geneticae Medicae et Gemellologiae* 20:96–113.

Terrenato, L., M. F. Gravina, A. San Martini, and L. Ulizzi. 1981. Natural selection associated with birth weight: III. Changes over the last twenty years. *Annals of Human Genetics* 45:267.

Thompson, Larry. 2000. Human gene therapy: Harsh lessons, high hopes. *FDA Consumer Magazine,* Sept.–Oct.

Thorndike, Edward L. 1905. Measurement of twins. *Archives of Philosophy, Psychology, and Scientific Methods,* no. 1 (Sept.): 1–64.

Tjio, J. H., and A. Levan. 1956a. The chromosome number of man. *American Journal of Obstetrics and Gynecology* 130:723–724.

―――. 1956b. The chromosome number of man. *Hereditas* 42:1–6.

Tjio, J. H., and T. T. Puck. 1958. The somatic chromosomes of man. *Proceedings of the National Academy of Sciences* 44:1229–1237.

Tjio, J. H., T. T. Puck, and A. Robinson. 1959. The somatic chromosomal constitution of some human subjects with genetic defects. *Proceedings of the National Academy of Sciences* 45:1008–1016.

Todes, Daniel P. 2002. *Pavlov's physiology factory: Experiment, interpretation, laboratory enterprise.* Baltimore: Johns Hopkins University Press.

Tufte, Edward R. 1997. *Visual explanations: Images and quantities, evidence and narrative.* Cheshire, CT: Graphics Press.

Vandenberg, Steven G., ed. 1968. *Progress in human behavior genetics: Recent reports on genetic syndromes, twin studies and statistical analyses.* Baltimore: Johns Hopkins University Press.

Vandenberg, Steven G., and John C. DeFries. 1970. Our hopes for behavior genetics. *Behavior Genetics* 1 (1): 1.

Vogel, F., and A. G. Motulsky. 1986. *Human genetics: Problems and approaches,* 2nd ed. Berlin: Springer-Verlag.

Vullo, R., and B. Modell. 1990. *What is thalassemia?* New York: Cooley's Anemia Foundation.

Wailoo, Keith. 1997. *Drawing blood: Technology and disease identity in twentieth century America.* Baltimore: Johns Hopkins University Press.

———. 2001. *Dying in the city of the blues: Sickle cell anemia and the politics of race and health.* Chapel Hill: University of North Carolina Press.

Warburton, Dorothy, D. A. Miller, O. J. Miller, P. W. Allderdice, and A. De Capoa. 1969. Detection of minute deletions in human karyotypes. *Cytogenetics* 8:97–108.

Watson, James. 1965. *The molecular biology of the gene.* New York: W. A. Benjamin.

Weatherall, D. J. 1991. *The new genetics and clinical practice.* Oxford: Oxford University Press.

Weindling, Paul. 1989. *Health, race and German politics between National Unification and Nazism, 1870–1945.* Cambridge: Cambridge University Press.

Weiner, Charles. 1994. Anticipating the consequences of genetic engineering: Past, present and future. In *Are genes us? The consequences of the new genetics,* ed. Carl F. Cranor, 31–51. New Brunswick, NJ: Rutgers University Press.

Weiss, Joan O., and Jayne S. Mackta. 1996. *Starting and sustaining genetic support groups.* Baltimore: Johns Hopkins University Press.

Weiss, Sheila Faith. 1987. *Race hygiene and national efficiency: The eugenics of Wilhelm Schallmayer.* Berkeley and Los Angeles: University of California Press.

Welton, Wooster, David Clayson, Felicia Axelrod, and David B. Levine. 1979. Intellectual development and familial dysautonomia. *Pediatrics* 63 (5): 708–712.

Wexler, Alice. 1995. *Mapping fate: A memoir of family, risk and genetic research.* Berkeley and Los Angeles: University of California Press.

Wise, Norton. 1995. *The values of precision.* Princeton: Princeton University Press.

Wolfe, S. M., and R. I. Henkin. 1970. Absence of taste in type II familial dysautonomia: unresponsiveness to methacholine despite the presence of taste buds. *Pediatrics* 77 (1): 103–108.

Woolf, Charles M., and Frank Dukepoo. 1959. Hopi Indians: Interbreeding and albinism. *Science* 164:30–37.

Woolf, L. I. 1966. Large-scale screening for metabolic disease in the newborn in Great Britain. In *Phenylketonuria and allied metabolic diseases,* ed. John A. Anderson and Kenneth F. Swaiman, 50–61. Washington, DC: U.S. Children's Bureau.

Woolf, Louis, and D. B. Vulliamy. 1951. Phenylketonuria with a study of the effect upon it of glutamic acid. *Archives of Disease in Childhood* 26:487–494.

World Health Organization. 1966. The use of twins in epidemiological studies (report of the WHO Meeting of Investigators on Methodology of Twin Studies). *Acta Geneticae Medicae et Gemellologiae* 15 (2): 113.

Wright, Susan. 1994. *Molecular politics: Developing American and British regulatory policy for genetic engineering.* Chicago: University of Chicago Press.

Yoxen, Edward. 1982. Constructing genetic diseases. In *The problem of medical knowledge: Examining the social construction of medicine,* ed. P. Wright and A. Treacher, 144–161. Edinburgh: Edinburgh University Press.

Zernig, Gerald, A. Saria, M. Kurz, and S. S. O'Malley. 2000. *Handbook of alcoholism.* Boca Raton: CRC Press.

Zihni, Lilian S. 1995. Mongolism, Down's syndrome and trisomy 21: Damning diagnoses in the twentieth century. *History and Philosophy of Psychology Newsletter* 21 (autumn): 20–25.

Zonderman, A. B. 1986. Twins, families, and the psychology of individual differences: The legacy of Steven G. Vandenberg. *Behavior Genetics* 16:11–24.

INDEX

Page numbers in *italics* refer to illustrations.

abortion, 15–16, 186, 202–3
achondroplasia, 66
Acta Geneticae Medicae et Gemellologiae (journal), 123
adenovirus vector, 196
alcoholism gene, 149–51
Allen, Richard J., 33–34
Alliance of Genetic Support Groups (later, Genetic Alliance), 164, 182, 200
Allison, Anthony C., 220n. 44
American Journal of Human Genetics, 11
American Journal of Medical Genetics, 202
American Society of Human Genetics, 10–11, 51–52, 53
Amish, Pennsylvania Old Order: "The Amish Madonna," 70; genealogies and, 61, 74; genetic diseases in, 60; Hostetler and, 67–68; Krusen and, 66; McKusick and, 66–72, 77–78; pedigree and, 63; as research collaborators, 4–5; as research subjects, 60–61, 68, 71–72; scientific papers and, 77–78

amniocentesis, 14–15
Anderson, John A., 50
Anderson, W. French, 189, 196, 197
animals: breeding of, 7, 121; genetic modification of, 208
Ashkenazi Jewish population, 157–58, 162, 163, 165, 186
Atomic Energy Commission, 194
Australian Twin Registry, 152
Axelrod, Felicia: clinical management by, 173–75; clinic at New York University and, 158; diagnosis and, 169–70; familiar seeing and, 172; papers by, 167, 168–69; prevention and, 175–76, 185–86

Bacillus subtilis ATCC 6051, 31
banding patterns, 118–19
Barnicot, N. A., 109
Barr, Murray, 14

Barton, Todd, 200
Bearn, Alexander, 51–52
Beebe, Gilbert, 133, 134, 138
behavior genetics, 1–2, 4, 121, 122–23, 153
Behavior Genetics Association, 123
Behavior Genetics (journal), 123
Bertram, E. G., 14
Bessman, Samuel P., 50–51
Bickel, Horst, 32
biochemical genetics, 1–2
bioethics industry, 198
biotechnology industry, 204
Bleuler, Eugen, 126
Bleyer, Adrien, 102
Block, Richard, 32
blood: disease as "in the blood," 7–9;
    genetic disease and, 54; testing for
    phenylketonuria (PKU), 31–34. *See
    also* screening programs
bodies: familial dysautonomia (FD)
    and, 161–62, 165–66, 168–69; as in-
    formation-retrieval systems, 54; as
    readouts of master text, 23; response
    to environment by, 191
breeding, 7, 121
British Alzheimer's Society, 207
Broder, Samuel, 189
Brunt, F. W., 162

Cambrosio, Alberto, 63
cancer: gene therapy for, 189; as ge-
    netic disease, 188–89, 190–91; litera-
    ture on, 190
cancer genetics, 1–2, 12, 99
Capecchi, Mario, 199
Caplan, Arthur, 198
*Cardiovascular Sound in Health and Dis-
    ease* (McKusick), 65
Carter-Saltzman, Louise, 223n. 1
Catcheside, D. G., 105
C-banding, 118

Centers for Disease Control, 94
centromeric index, 104
Chalmers, Thomas, 141
Chapelle, Albert de la, 109
Childs, Barton, 9, 10, 14, 52, 191
chromosomal location, 80–81
chromosomes: ambiguity of, 119; cyto-
    genetic analysis of, 100–102; disease
    and, 92, 102–5, 115; measuring of,
    106; naming and classification of,
    104–12; numbering of, 91; Philadel-
    phia, 12; prenatal diagnosis and, 94;
    seeing, 95–102; as "squashed spi-
    ders," 90; visual images of, 115
"Chromosomes for Beginners"
    (Lennox), 90
Chu, Ernest, 104, 116
Clayson, David, 167, 168–69
Cline, Martin, 195
Cohen, Bernard, 137–38
Collins, Francis: on cancer genetics,
    190; on disease, 17; Guttmacher and,
    201; at People's Genome Celebra-
    tion, 200, 206; as promoter of
    genome project, 18
Columbia University, 113–15
constitutional illness, 7–9, 11–12, 17
consumer: expertise of, 182; role of in
    biomedical science, 207, 210
Corey, Linda A., 152
Corner House, 207
corporate culture and biology, 181, 196,
    205–6
Cotterman, Charles, 11
craft knowledge, 4, 62, 221–22n. 5
Crow, James F., 51
cytogenetics: attraction to, 12; begin-
    ning of, 98; chromosomes and,
    100–102; complacency in field of,
    98; familiar seeing and, 112–15; ge-
    netic disease and, 115–16; interna-

tional standards for, 110–12; Lennox on, 91; transformation of, 1–2

Dancis, Joseph, 15, 170–71, 173
Danish Twin Registry, 131
Darwin, Charles, 7, 122
Davens, Edward, 48
Day, Richard, 157
DeBakey, Michael E., 132–33, 141
Denver Conference on Nomenclature, 91, 94, 105–12, 119
Department of Health, Education and Welfare. *See* U.S. Children's Bureau
disease: all human, as genetic, 2, 17–18, 23–24, 188, 191–92; as biosocial experience, 192–93; cataloguing, 79–85; chromosomes and, 92, 102–5, 115; injustice and, 25–26; as "in the blood," 7–9; legislation specific to, 49; social organization, technology, and, 157; testing technologies and, 39; timing of diagnosis of, 36–37; view of by clinicians, 11–12. *See also* genetic disease; *specific diseases*
DNA, 6–7, 27
DNA-chip technology, 18, 57
Dobbs, June M., 44
Dor Yeshorim, 181
Down, John Langdon, 8
Down syndrome, 8, 103, 106–8
Dunn, L. C., 140–41
Dutch newborn, skeleton of, 78–79, 79
dwarfing conditions, 59, 66, 69
"Dwarfism in the Amish" (McKusick), 75, 78
Dysautonomia Association, Inc. (later Dysautonomia Foundation), 164–67, 178, 184
Dysautonomia Day, 175, 176
Dysautonomia Treatment and Evaluation Center, 174, 175

earnings and innate ability, 144–47
Egeland, Janice, 69, 84
Ehrlich, Paul, *Population Bomb*, 15
Eldridge, Roswell, 84
Ellis–van Creveld (EVC) syndrome, 59, 69, 70
emotional knowledge, 4, 62, 179
English Men of Science (Galton), 124
environmentalists, 206–7
ethics: bioethics industry, 198; phenylketonuria (PKU) and, 41–42, 43–44, 55
eugenics movement, 8–9, 15, 208–9

face, seeing genetic disease in, 221–22n. 5
familial dysautonomia (FD): clinical management of, 173–75; clinical visibility of, 156–57, 160–61; consensus regarding, 158–59; diagnosis of, 169–72; familiar seeing and, 172; gastrostomy and fundoplication strategy in, 174–75; gene for, 158, 176–82; as genetic disease, 159; as hereditary sensory and autonomic neuropathy (HSAN), 172; heredity and, 157–58, 161–63; intellectual and psychological development in, 167–69; parent support group and, 164–67; pregnancy and, 187; prevention of, 175–76, 185–87; publicity about, 187; symptoms of, 160–61, 185; tocotrienol and, 183–84. *See also* Axelrod, Felicia
familiar seeing, 86, 112–15, 172
Fanconi Anemia Research Fund, Inc., 177
FD. *See* familial dysautonomia (FD)
FDA, 197, 198
FD Hope, 181–82
Feinleib, Manning, 141

field collection and data analysis: Allison and, 220n. 44; in biomedical research, 59–60; historical reconstruction and, 87–88; McKusick and, 62–63; postwar genetics and, 88; properties of, 85

field trials: of Guthrie test, 36, 37–38; of intradermal histamine test, 170–71

filter paper, soaking through, 35–36

fingerprints and zygosity, 138, 140

folk epidemiology, 73–74

folk knowledge: behavior genetics and, 4, 121; determination of zygosity and, 121; familial dysautonomia (FD) and, 184; genetic disease and, 205; of heredity, 7; of human nature, 153–54; pedigree and, 62; technical knowledge and, 24–25

Folling, Asbjorn, 31–32

Ford, Charles E., 98, 100, 104, 105, 109

founder effect, 162

Frankel, Marjorie Eustis, 165

Frankensalmon, 208

Fraser, Clarke, 101

Fraumeni, Joseph, 191

Freeman, Alfred, 165

Freeman, Frank, 127–30

galactosemia, 41, 56

Galton, Francis, 122, 124–25

Garrod, Archibald, *Inborn Errors of Metabolism*, 9–10

Gautier, Marthe, 103

Gedda, Luigi, 123

Gelsinger, Jesse, 195, 196, 197

gene, discovery of, 158, 176–82, 200–202

gene frequencies, 17–18

gene therapy: "adverse events" and, 195–96; cancer and, 189; as disappointment, 197–98, 199; early

enthusiasm for, 189–90, 197; for hemophilia, 198–99; types of, 194–95; viral vectors and, 197

Genetic Alliance (formerly, Alliance of Genetic Support Groups), 164, 182, 200

genetic disease: American Society of Human Genetics and, 53; blood and, 54; chromosomal location and, 80–81; clinical care for, 173; conditions mimicking, 84; cytogenetics and, 115–16; emphasis on, 24–26; fortunes of, 16–17; "hidden cost" of, 217n. 44; legislative initiatives and, 53–54; as public health problem, 55, 56; qualities of, 27; rise to medical prominence of, 52, 54; seeing in faces, 221–22n. 5; as simplifying health problems, 199–200; as small, 156; technological optimism and, 17; transformations in, 18–20, 184–85; twin studies and, 135–36. *See also specific diseases*

Genetic Diseases Act, 182

genetic engineering, 193, 208

"Genetic Studies of the Amish: Background and Potentialities" (McKusick, Hostetler, & Egeland), 69, 71

genetic testing, 182, 201–2. *See also* prenatal diagnosis

Gene Transfer Safety Symposia, 198

Genomic Health, 178

genomic medicine, 204, 205–6, 209–10

genomic research and commercial networks, 181, 196

Genovo, 196

Genzyme Genetics, 182

germ-line gene therapy, 194–95, 199, 209

Giannelli, F., 116–17

Giemsa banding, 118

Ginsburg, Faye, 182–83, 187

Glass, Bentley, 15, 88
Gordon, Hymie, 13
Gottesman, Michael, 189–90
Gregg, Alan, 122, 147
Guidotti, Rick, 200
Gusella, James, 158, 178, 179, 180
Guthrie, Robert, 31, 32–33, 37
Guttmacher, Alan, 201

Hamerton, John L., 94, 98, 100, 109
Hansemann, D., 95
Haraway, Donna, 19
Harris, Harry, *Human Biochemical Genetics*, 10
Harvard University, 158, 178, 179, 180
Hayes, Richard, 206–7
health care delivery system and bodily fluids, 32–33
heel prick for neonatal testing, 35
hemophilia, 8, 198–99
hereditary sensory and autonomic neuropathies (HSANs), 171–72
heredity: constitutional illness and, 8; disease and, 26; earnings and, 145–47; environmental modifications and, 13–14; eugenics movement and, 8–9; familial dysautonomia (FD) and, 157–58, 161–63; folk knowledge of, 7; "instinctual" behaviors and, 122; intelligence and, 122
Heritable Disorders of Connective Tissue (McKusick), 65
Herndon, C. Nash, 11
heterozygote advantage hypothesis, 162–63
Holzinger, Karl J., 127–30
Hood, Leroy, 199
Hormuth, Rudolph P., 38
Hostetler, John, 67–69, 84
Howlett, R. M., 116–17
Hrubec, Zdenek, 141, 143, 146

HSANs (hereditary sensory and autonomic neuropathies), 171–72
Hsu, T. C.: chromosomes and, 104; on complacency of field, 98–99; Denver Conference and, 94, 105; on Down syndrome, 101; on squash techniques, 100
Human Biochemical Genetics (Harris), 10
human cloning, 17
human genetics: as career, 10–11; field research and, 87–88; parent support group and, 184; phenylketonuria (PKU) and, 30, 51–54; postwar, 59
human genome, mapping of, 193–95
Human Genome Organization, 194
Human Genome Project: abortion and, 202; cataloguing and, 89; genetic disease and, 24; genetic testing and, 57; McKusick and, 88; *Mendelian Inheritance in Man* and, 80, 81
human karyotypes, 14, 94, 109–10, 113, 116–18
human subjects: Amish as, 60–61, 68, 71–72; cooperation of, 101, 128–29; death of, 195, 196, 197; in experimental system, 3; gene therapy and, 197–98; information on, 116; rules for referring to, 111; twins as, 140–44; veterans as, 133–34
Hungerford, David, 12, 105
Huntington, George, 8
Huntington disease, 8
hypotonic solution, 98–99

idiogram, 97, 104
Inborn Errors of Metabolism (Garrod), 9–10
infant mortality, causes of, 13–14
informed consent, 197
Inquiries into Human Faculty and its Development (Galton), 125

Institute for Human Gene Therapy, University of Pennsylvania, 195
International Association of Biological Standards, 93
International Classification of Diseases, World Health Organization, 111
International Society for Twin Studies, 152
intradermal histamine reaction, 169, 170–71
involuntary sterilization, 8, 126–27

Jablon, Seymour, 140
Jacobs, Patricia, 104, 105, 109
Jensen, Arthur, 147, 223–24n. 6
Jervis, George, 32
Johns Hopkins University: Medical School, Division of Medical Genetics, 65; Prenatal Birth Defects Center, 16

Kaij, Lennart, 150
Kallmann, Franz, 126–27
Kelley, William N., 196
Kirk, Norman, 132–33
knowledge: biomedical and community, 6; of disability, segregation of, 182–83; familiar seeing, 86, 112–15; family support groups and, 164–65, 166; genetic testing and, 57; genomic medicine and, 24–25; increased, 26–27; of natural phenomena, 6; physical location of, 183; as socially produced, 205; types of, 4–5. See also specific types of knowledge
Kohler, Robert, 85
Kraepelin, Emil, 126
Krueger, Gladys, 36, 42, 43
Krusen, David E., 66, 68
Kuklick, Henrika, 85

Lajtha, L. G., 104
Lancet (journal), 90

Lange, Johannes, 131
Lederberg, Joshua, 51
Lejeune, Jerome: chromosomes and, 104; Denver Conference and, 105; Down syndrome and, 10, 101, 103, 106–7; London conference and, 109
Lemkau, Paul, 142
Lennox, Bernard, 90–91
Lenz, Fritz, 126–27
Levan, Albert, 97, 98–100, 104, 105
Living with a Child with Familial Dysautonomia (pamphlet, Riley), 166–67
Lofenalac, 41
London Conference on the Normal Human Karyotype, 109–10
Luxenburger, H., 130–31

Macklin, Madge, 11
Manual of Comprehensive Care (Axelrod), 173
Marfan syndrome, 66
Margus, Brad, 201
Marsh, Lucille J., 37–38, 39
Martin, Ruby G., 38
Maryland: PKU screening legislation in, 46–49; state office of vital statistics in, 136–37
Massachusetts General Hospital, 176, 180, 181
McKusick, Victor: Amish and, 4–5, 66–72, 77–78; Bar Harbor short course of, 10, 13; career of, 64–65; database of, 74–75; Ellis–van Creveld syndrome and, 61; familial dysautonomia (FD) and, 162; field methods of, 62–63, 66, 68–69, 75–79; genealogies and, 61; Hostetler and, 67; Mendelian Inheritance in Man, 20, 79–85; network of, 72–75; pedigree methods and, 85–86; science, medicine, and, 86–87; Tangier Island and,

65–66; Vanilla and, 87; works of, 77–78, 88

McKusick, Vincent, 76

McMahan, Brian, 141

Medical Follow-Up Agency, 132, 134, 135

Melby, Edward C., 87

Mendelian Inheritance in Man (McKusick), 20, 79–85

mental retardation, 40–41, 218n. 18. *See also* Down syndrome

methodology, 7

metrology, 92

Meyers, Abby S., 199

Mid-Atlantic Twin Registry, 152

Miles Laboratories, 39

Milunsky, Aubrey, 16

"moment of truth": description of, 2–3; in genomic medicine, 209–10; of machine, 112; pedigree and, 62–63

Morse, Robert, 146

Motulsky, Arno, 88

Muench, Karl H., 24

Muller, H. J., 10, 11, 105, 127–30

Murray, Robert F., 52

Myers, Fred, 187

Myers, Samantha, 187

Nance, Walter E., 152

National Academy of Sciences Current Era Twin Registry, 151–52

National Academy of Sciences— National Research Council Veteran Twin Registry: creation of, 132–34; description of, 121; finding twins for, 134–38; uses of, 140–44; zygosity determination and, 138–40

National Association for Retarded Children (NARC), 45–46

National Institutes of Health: funding for FD research by, 176; Human Gene Therapy Working Group, 195;

investigation by, 197; Office of Genome Research, 194; Recombinant DNA Advisory Committee, 189

National Organization for Rare Disorders, 199

National Research Council, 146. *See also* National Academy of Sciences— National Research Council Veteran Twin Registry

National Twin Registry of Sri Lanka, 152

Nature Genetics, 198–99

Neel, James V.: Allen and, 33; career of, 11; field collection and, 60; phenylketonuria (PKU) and, 51; twin registry and, 135–36, 138, 141; works of, 87–88

Nelkin, Dorothy, 6–7

Newman, Horatio, 127–30

New York University, 158, 174, 175

North Carolina Twin Registry, 152

Northern California Twin Registry, 152

Norton, Horace, 11

Nowell, Peter, 12

Nukaga, Yoshio, 63

obesity, 203–4

Odesina, Victoria, 200–201

ornithine transcarbamylase deficiency, 196

Painter, T. S., 95–96, 96, 116

parental knowledge: Allen and, 33–34; familial dysautonomia (FD) and, 5, 164–65; Guthrie and, 33; McKusick and, 73–74; scientific texts and, 153, 178

parent support group: Dysautonomia Association, Inc., 164–67, 178, 184; funding for research and, 165, 177, 201; human genetics and, 184; as network for production of knowledge, 166; splintering of, 179–80, 181–82

Patau, Klaus, 91, 107, 108–9
patchwork quilt analogy of scientific
    texts, 5–6
patent for gene, 181
Paul, Diane, 29, 41
peas, image of, 121–22
pedigree, 62–63, 85–86, 87
Peele, Stanton, 151
Penrose, Lionel, 32, 37, 109, 116, 117–18
People's Genome Celebration,
    200–201, 206
pharmacogenomics, 18
phenotype, 81, 84, 111
phenylketonuria (PKU): bacterial
    growth halo in positive test of, 34;
    blood test for, 31–34; description of,
    28; dietary intervention for, 12–13,
    29, 40–45; doubts about program
    for, 49–51; geneticists and, 51–54; in-
    hibition assay for, 34–37; legislative
    action and, 45–49; as local historical
    moment, 55–56; materials given to
    new mothers about, 40; medical and
    political management system and,
    28–30; as public health problem, 55;
    as racial disease, 37–40; screening
    program for, 54–55; variability of
    course of, 44, 50
Philadelphia chromosome, 12
phytohemagglutinin, 90
Ploetz, Alfred, 126
Population Bomb (Ehrlich), 15
populations, "primitive," 59–60. *See
    also* Amish, Pennsylvania Old Order
Porter, Ted, 93
Prather, Perry, 48
preimplantation diagnosis and selec-
    tion, 18
prenatal diagnosis: amniocentesis and,
    14–15; chromosomes and, 94; ge-
    netic disease and, 186

prenatal testing, 182–83. *See also*
    genetic testing
Price, Bronson, 11
privacy of patients, 111, 141, 142
pseudoxanthoma elasticum, 178
psychiatric genetics, 126
public health genetics, 1–2, 12–13
Puck, Theodore T., 104–5, 109

quinacrine dihydrochloride (Q-band-
    ing), 118

radiation damage, 194
Rapp, Rayna, 182–83
Riley, Conrad, 157, 160, 165–67
Riley-Day syndrome, 157. *See also* famil-
    ial dysautonomia (FD)
Rockefeller Foundation, 122
Roe v. Wade, 14, 16
Roney, J. Albert, Jr., 47–48
Roseman, Lenore, 179, 229n. 23
Rosenberg, Charles, 8
Rosenberg, Steven, 189
Rubin, Berish, 179, 180–81, 183–84
Rubinstein, Jack H., 44
Rudin, Ernst, 126

Salzano, Francisco, 59
Scarr, Sandra, 153, 223n. 1
Schaffer, Simon, 92–93
Schilder's disease, 83
schizophrenia, 83, 126–27, 142
Schull, William, 88
scientific texts: incorporation of conclu-
    sions of non-technically trained
    people into, 3–5; parental knowledge
    and, 153, 164–65, 178; parent support
    group and, 184; patchwork quilt
    analogy of, 5–6; ways of seeing, 3
Scott, Randy, 178
screening programs: contention over,

54–55, 56; for familial dysautonomia (FD), 156–57, 158, 169, 170–71; gene discovery and, 201–2; legislative action for, 45–49; for phenylketonuria (PKU), 28–30, 36, 37–38; rise of genetic disease to medical prominence and, 52; tandem mass spectrometry and, 56

sex chromatin, 14

Sexton, Sarah, 207

sickle cell anemia, 163, 200–201

Sickle Cell Anemia Control Act, 53–54

Siemens, Hermann, 125

Silver, Lee, 18, 19, 199

Slaugenhaupt, Susan, 176, 177, 180, 181

Smith, Alfred A., 170–71

Smocovitis, Vassiliki Betty, 88

social knowledge: behavior genetics and, 4, 121; determination of zygosity and, 121; from living with disease, 183; pedigree and, 62

somatic gene therapy, 195

Spuhler, J. N., 123

squash techniques for cells, 99–100

staining techniques, 118–19

standardization: of banding patterns, 118–19; cytogenetics and, 110–12; definition of, 92; Denver Conference on Nomenclature and, 105–12; disciplinary and historical elements of, 117; discussion of, 94–95; international pressure for, 94; Penrose on, 118; practices of, 92–93

starvation gene, 204

Stern, Curt, 105, 109

Study of Inborn Errors of Metabolism, 52–53

Sulston, John, 25

Susi, Ada, 32, 33

Swaiman, Kenneth F., 50

tandem mass spectrometry, 56

Tangier Island, 65–66

Taubman, Paul, 144–47

Taussig, Karen-Sue, 178

Tay-Sachs disease, 159

technical knowledge: determination of zygosity and, 121; folk knowledge and, 24–25; genetic disease and, 205; judgment and, 112; pedigree and, 62; process of producing, 3

technological optimism, 17, 206

Terry, Patrick and Sharon, 178

thalassemia, 54, 195

Tijo, Joe-Hin, 97, 98–100, 104, 105, 109

tocotrienol, 183–84

Tourette, Georges Gilles de la, 8

Tourette syndrome, 8

trisomy 21, 106–8

Tufte, Edward R., 64

Turpin, Raymond, 103, 104

"twinning," phenomenon of, 130

twin registries, 123–24, 130–32, 151–52. See also National Academy of Sciences—National Research Council Veteran Twin Registry

twin studies: alcoholism gene and, 149–51; "earnings" and, 144–47; Galton and, 122, 124–25; in Germany, 125; intelligence and, 147; by Newman, Freeman, and Holzinger, 127–30; reports to twins on, 143; schizophrenia and, 126–27, 142; self as changeless and, 153–55; "twin method" and, 125; zygosity and, 4, 120–24, 138–40. See also twin registries

U.S. Children's Bureau: Collaborative Project of, 45, 47; genetic disease and, 182; PKU dietary intervention and, 42–43; PKU field testing and, 36, 37, 38–39

U.S. Department of Energy, 194
U.S. Public Health Service, 39
University of Chicago, 127
University of Pennsylvania: bioethics
    program, 198; Institute for Human
    Gene Therapy, 195, 197, 198; Medical
    Center, 196

Varmus, Harold, 197
Virginia Twin Registry, 152
vital statistics, state offices of, 136–38
Vogelstein, Bert, 188–89

Waardenburg, P. J., 102
Walker, Norma Ford, 11
Warburton, Dorothy, 113–15
Watson, James, 18, 190, 194
Watson, John Broadus, 153
Weill, Maryon, 229n. 23

Weiner, Alexander, 11
Welch, William, 33
Wellcome Trust, 207
Welton, Wooster, 167, 168–69
White, Benjamin, 47
"wife effects," 146
Wilson, James, 195, 196, 198, 199
Winiwarter, Hans von, 95
World Health Organization, 93, 111
Wright, Sewall, 223–24n. 6

X chromosome, traits on, 82
xeroderma pigmentosum, 8

Yoxen, Edward, 12

zygosity, determination of, 4, 120–21,
    138–40